Water Pollution Control

Ralph A. Luken
Edward H. Pechan

andrew

The Praeger Special Studies program—
utilizing the most modern and efficient book
production techniques and a selective
worldwide distribution network—makes
available to the academic, government, and
business communities significant, timely
research in U.S. and international economic, social, and political development.

Water Pollution Control

Assessing the Impacts and Costs of Environmental Standards

PRAEGER SPECIAL STUDIES • DESIGN/ENVIRONMENTAL PLANNING SERIES

Praeger Publishers   New York   London

Library of Congress Cataloging in Publication Data

Luken, Ralph Andrew.
    Water Pollution control.

    (Praeger special studies, design/environmental
planning series)
    Includes bibliographical references.
    1. Water quality management—United States.
2. Water quality management—United States—Cost
effectiveness.   3. Water—Pollution—Law and
legislation—United States.   I. Pechan, Edward H.,
1947-   joint author.   II. Title.
TD223. L77   363.6'1   77-3071
ISBN 0-275-24470-9

PRAEGER PUBLISHERS
200 Park Avenue, New York, N.Y. 10017, U.S.A.

Published in the United States of America in 1977
by Praeger Publishers, Inc.

789   038   987654321

Printed in the United States of America

As concerns over the problems of water pollution in the United States have changed over time, so have the policies for control of pollution. The trend has been one of increasing public concern over the environmental degradation caused by unchecked water pollution, a trend which has led to an increasing federal government responsibility. This book addresses one of the key policy changes enacted into law by the Congress in 1972—that of requiring certain categories of industries and municipalities to meet pollution discharge standards based on specific technology objectives regardless of the prior conditions or desired uses of the water.

The procedures used here to address this issue consist of (1) design of a conceptual approach to provide quantitative information concerning the issue based on available information, (2) exercise of the quantitative analysis tools to examine both current policies and some logical alternatives, (3) analysis of the quantitative results in order to generate findings and conclusions, and (4) consideration of the findings and conclusions in order to gain insights into potential alternatives that are both more efficient than current policies and are practicable from the standpoint of implementation. These insights are presented in the final chapter of the book as specific recommendations for a change in the federal water pollution control legislation.

While the quantitative analyses presented here are applicable only in the analysis of water pollution problems, the overall method of developing, executing, and using results based on the study framework is generically applicable to the study of other major public policy issues. Thus, this book presents a framework for problem study with wide application as well as a set of specific results.

ACKNOWLEDGMENTS

Much of the basic work for this study was supported by the National Academy of Sciences/National Research Council (NAS/NRC) as part of a contract with the National Commission on Water Quality. Mr. Theodore M. Schad, Executive Secretary of the Environmental Studies Board, was especially helpful in his encouragement and support of the project.

Several members of the NAS/NRC Study Committee on Water Quality Policy made key contributions. Mr. M. Gordon Wolman, Chairman of the Committee, provided overall aid. Blair T. Bower provided substantive guidance in developing the study concept. Specific assistance was provided by Drs. Perry McCarty, Neal Armstrong, and Howard Madsden on cost analysis, water quality, and non-point sources respectively. Many NAS/NRC staff members contributed to the project. Daniel Basta contributed to the analysis in most of the chapters of the report submitted to the Study Committee on Water Quality Policy. Other staff members deserving mention for their effort are Forest Arnold and Larry Wallace.

CONTENTS

LIST OF TABLES

xi

LIST OF FIGURES

LIST OF ACRONYMS

BAT   Best available technology

BOD   Biochemical oxygen demand

BPT   Best practicable technology

BPWTT   Best practicable waste treatment technology

COD   Chemical oxygen demand

EPA   Environmental Protection Agency

GAO   General Accounting Office

MRI   Midwest Research Institute

N   Nitrogen

NCWQ   National Commission on Water Quality

NRDI   National residuals discharge inventory

NSPS   New Source performance standards

OBERS   Office of Business Economics and Economic Research
           Service

P   Phosphorus

SIC   Standard industrial classification

SMSA   Standard Metropolitan statistical area

SS   Suspended solids

ST   Secondary treatment

WRC   Water Resources Council

Water Pollution Control

1

INTRODUCTION

FEDERAL WATER POLLUTION CONTROL ACT AMENDMENTS OF 1972

The Federal Water Pollution Control Act Amendments of 1972 (hereafter referred to as Public Law [PL] 92-500, the 1972 Act or the 1972 Law) marked a decisive shift in the nation's approach to restoring and maintaining the physical, chemical, and biological integrity of its waters (Public Law 92-500 1972). That shift is best reflected by the major change in control mechanisms. Under the Federal Water Pollution Act Amendments of 1965 (hereafter referred to as the 1965 Act or 1965 Law), water quality standards for the nation's waters were set as the control mechanism. The use of the waters for such activities as drinking, recreation, and manufacturing determined the kinds and amounts of residuals to be discharged, the degree of residual abatement required, and the promptness with which dischargers were to install the necessary abatement technology. Under the 1972 Act, effluent limitations were set as the control mechanism. The existence and availability of water pollution control technology determines the kinds and amounts of residuals to be discharged, and legislatively mandated compliance dates determine the promptness with which dischargers must install the necessary abatement technology.

The overwhelming congressional support for the 1972 Law resulted from public disillusionment with the lack of progress in water pollution control after more than two decades of federal legislation. The 1970 and 1971 hearings of the Senate Subcommittee on Air and Water Pollution catalogued the legislation's major limitations, some of which are pertinent to this study (Library of Congress 1973).

First, the setting of water quality standards by the states
for interstate navigable waters did not immediately become the
envisaged keystone for an effective program. The water quality
standards were expected to establish the maximum level of degradation allowable and provide an avenue for legal action against
polluters. However, only a little more than half of the states had
fully approved standards four years after the deadline for their
submission. The lack of approved standards also delayed federal
enforcement actions; before beginning a legal action to abate pollution, the federal government had to wait for a noticeable violation
of water quality—for pollution originating in one state to endanger
the health or welfare of persons in another state—and for consent
of the governor of the state in which pollution originated.

Second, the abatement procedure established in the original
Federal Water Pollution Control Act, passed in 1948, and carried
forward in the 1965 Law contributed to the delay in enforcement
actions against polluters. The 1948 abatement procedure provided
for conferences and negotiations between dischargers of pollutants
and government and allowed a court to order abatement only after
finding that compliance with the order was feasible. Under this
procedure, only one case reached the courts in more than two
decades. Even in that case, the city involved ended up treating
only half of its sewage within two years of completing the court-
mandated sewage treatment plant.

Third, the failure to fully implement the federal construction
grant program for municipal sewage treatment plants delayed
action by municipalities to build the necessary treatment facilities.
Of the $3.4 billion authorized for this purpose by the 1965 legisla-
tion, only $2.2 billion was appropriated. Even the availability of
the full $3.4 billion would probably not have accelerated completion
of the needed facilities, because the total cost of needed facilities
was estimated to be between $20 and $30 billion, and the federal
grant at the maximum level would cover only 55 percent of the
capital costs and none of the operation and maintenance costs in-
curred by local governments.

Fourth, the federal government did not have adequate authority
to collect the data on industrial dischargers necessary to administer
an effective program. The 1965 Act did permit the Environmental
Protection Agency (EPA) and its predecessors to require the filing
of reports by industrial and municipal dischargers after an enforcement conference. The reports were to contain data on discharges
and proposed abatement actions. However, a General Accounting
Office (GAO) investigation of state permits issued to 80 industrial
plants discharging into a 170-mile reach of the Mississippi River

found that there were data on waste discharge from only 52 of the 82 plants. More important than the absence of data on these individual plants was the fact that dischargers were allowed to omit from their reports any information they believed to involve trade secrets, and that EPA had no legal right of entry to check dischargers' operations.

The 1972 Act attempted to respond to these limitations with three essential elements: uniformity, finality, and enforceability. In the words of the law's author, Senator Edmund Muskie: "Without these elements, a new law [1972 Law] would not constitute any improvement on the old [1965 Law]; we would not bring a conference agreement to the floor without them" (Library of Congress 1973).

Uniformity is mandated by the requirement that each residual discharger within a category or class of industrial sources, and all municipal sources, must meet stipulated effluent limitations regardless of geographic location. Each category or class of industrial sources is required to meet nationally uniform effluent limitations based on best practicable (control) technology (BPT) currently available by 1977, and to meet even more stringent nationally uniform effluent limitations based on best available technology (BAT) economically achievable by 1983. For example, all plants in the ironmaking subcategory of the iron and steel industry, regardless of location, are required to meet effluent limitations defined in pounds of residual per ton of product output for such parameters as suspended and dissolved solids. All municipal sources (publicly owned treatment works) are required to meet effluent limitations based on secondary treatment (ST) by 1977 and to meet even more stringent effluent limitations based on best practicable waste treatment technology (BPWTT) by 1983.

Finality is mandated by the requirement that point-source dischargers (municipal and industrial activities) meet more stringent effluent limitations at specific dates in the future. The goal of the 1965 Act was to maintain and restore the quality of the nation's water without mandating specific pollution control activities and without setting a specific date for meeting that goal. The 1972 Act requires dischargers to meet one set of effluent limitations in 1977, a more stringent set of effluent limitations in 1983, and looks toward achieving the final goal of zero discharge of pollutants into any navigable waters by 1985. In addition, an interim goal of achieving waters fit for fishing and swimming by 1983 is established. The concept of finality is intended to remove the uncertainty on the part of industrial and municipal dischargers about the nation's (or at least Congress's) commitment to maintain and restore the quality of the nation's waters. The 1972 Act attempts to give point-source residual dischargers a specifically stated period during which they will be guaranteed that effluent limitations will not change and they will know the date at which they will be required to meet the effluent limitations.

Enforceability is assured through the provisions of the discharger permit program and the new enforcement authorities given to EPA. The 1972 Act is based on the assumption that violations of permit conditions would be easier to determine than violations of water quality standards, assuming that the government has the ability to design an adequate compliance monitoring program and to inspect the operations of residual dischargers. EPA has the authority and is required to issue an abatement order whenever there is a violation of the terms or conditions of a permit and a state fails to move against a violator in a timely fashion. Furthermore, a citizen may bring suit against EPA if it fails to issue a necessary order.

Although the 1972 Act in its final form received congressional support sufficient to override a presidential veto, there were many compromises in the development of the final version as it moved through the procedures of the Senate and the House of Representatives and of the Conference Committee created to resolve the differences between versions of the legislation passed by the Senate and the House. Many members of Congress doubted the wisdom of mandating increasingly more stringent, uniform effluent limitations leading up to a final commitment to the elimination of all discharges of pollutants into water. Thus, while the three major innovative provisions of the Senate version survived in the legislation agreed to by the Conference Committee, a provision was inserted that established a national study commission, under Section 315, to "make a full and complete investigation and study of all of the technological aspects of achieving, and all aspects of the total economic, social, and environmental effects of achieving or not achieving the effluent limitations and goals set forth for 1983" (Public Law 92-500 1972), looking toward recommendations not later than October 18, 1975 as to any needed "mid-course corrections that may be necessary" (Library of Congress 1973).

NATIONAL COMMISSION ON WATER QUALITY

The National Commission on Water Quality (NCWQ) thus created under Section 315 initiated in the summer of 1973 its study of the consequences of achieving or not achieving the goals of the act. The NCWQ defined five major areas of study to provide data for its evaluation of the effects of the 1972 Act: technological assessment, economic and social impact, water quality analysis and environmental impact assessment, institutional assessment, and regional studies. A report including the recommendations of the commissioners was transmitted to Congress in March 1976, and a staff report covering these five areas, together with the staff's findings, was released in April 1976.

A brief summary of the findings and recommendations of the NCWQ is given in Appendix A. Specific findings and recommendations are discussed throughout the book whenever they are relevant.

In implementing its study program, the NCWQ appeared to accept the uniformity provisions of the act, and concentrated its contract studies primarily on the finality provisions and secondarily on the enforceability provisions. The NCWQ's contractual studies covered essentially an analysis of the availability and cost of end-of-pipe technology to meet requirements for control of industrial and municipal point-source discharges; and the social, economic, and environmental consequences of meeting the 1977 and 1983 effluent limitations and the 1985 goal of elimination of discharge. These studies provided data to determine whether the finality provisions are feasible and whether the benefits are commensurate with the costs. Another set of NCWQ studies contained analyses of the institutional arrangements provided under the 1972 Act. The institutional studies, broadly interpreted, provided a basis for determining whether the finality provisions of the act are enforceable. These studies covered not only the permit program, compliance monitoring, and enforcement provisions, but also activities such as planning and grant provisions, which, if successful, would eliminate the need for enforcement actions. The regional studies analyzed effects of the act in a number of specific geographic locations, but the studies did not question in most cases the benefits or costs of requiring uniform treatment of similar classes of residual discharges regardless of geographic location. (While the act does allow state and federal regulatory agencies to require controls more stringent than the technologically defined effluent limitations where a greater degree of treatment is needed to achieve water quality standards, it does not allow for less stringent effluent limitations if they are not needed to meet water quality standards.)

ASSISTANCE FROM THE NATIONAL ACADEMY OF SCIENCES

Early in the course of its study program, the NCWQ contracted for assistance from the National Research Council of the

National Academy of Sciences, under the provisions of Section 315 of the act.

In order to provide the assistance needed, the Environmental Studies Board of the National Research Council created the Study Committee on Water Quality Policy. In connection with its accomplishment of the tasks assigned by the NCWQ, the committee determined that an independent assessment of residuals generation and discharge, and of the cost of residuals abatement technologies, was essential to provide perspective on its assignment. In the absence of a breakdown by NCWQ of national totals by geographic regions, the committee engaged consultants and directed them to devise a system that could provide it with a basis for an independent analysis of the effects of achieving or not achieving the goals of the act. This analysis provided a basis for the committee's comments on the NCWQ's contractor and staff draft reports.

The result of the consultants' effort was preparation of the national residuals discharge inventory (NRDI) (Luken, Basta, and Pechan 1976). The NRDI contains a quantitative assessment of residuals* under both gross (uncontrolled) and net (controlled) conditions in each of the 3,111 counties in the contiguous United States. The NRDI analyses include point-source discharges, which are defined as discharges from municipal and industrial activities, and areal-source discharges, which are defined as urban runoff and runoff from nonirrigated agricultural activities.

The NRDI report to the committee described the compilation techniques of NRDI and the data sources used, and developed findings on the following issues:

1. The distribution of the diverse residuals-generating activities and types of waterborne residuals in all river basins constituting the continental United States, and the implications of this diversity;

2. The relative effects on a simple measure of water quality of applying BPT/ST and BAT/BPWTT to industrial and municipal activities is each of the river basins;

3. The estimated costs to the nation of applying BPT/ST and BAT/BPWTT to industrial and municipal activities; and

4. The estimated costs to the nation of pursuing policies other than the imposition of national uniform effluent limitations on industrial and municipal activities.

---

*"A residual is a quantity of material or energy left over after inputs are converted into outputs." In common parlance, residuals are called pollutants. NCWQ referred to them as effluents.

This book is based on the report to the committee and is written by two of the three consultants who prepared the report. The book is different from the NRDI report in that (1) it is an explicit examination of alternatives to the uniformity policy; (2) it incorporates additional material including legislative history of the 1972 Act, principal NCWQ findings and recommendations, some EPA reactions to the NCWQ recommendations, additional policy alternatives, findings, and conclusions based on the NRDI analysis, and a recommendation for modification of the 1972 Law; (3) it corrects many of the mistakes in the NRDI report, which resulted from the pressure of preparing such a comprehensive report in a short time; and (4) it simplifies the presentation of technical material to make it more readable.

INEFFICIENCIES OF UNIFORMITY

While the recommendations of the NCWQ reflect the need for some flexibility in the 1972 Act, the findings of the NCWQ did not explicitly document the costs or environmental consequences of alternative policies. The NCWQ's findings focused on the finality and enforceability provisions because its mandate was to examine the consequences of achieving or not achieving the goals (primarily technological) of the act. NCWQ was not requested to analyze the consequences of alternative policies and thus only tangentially examined the consequences of the uniformity provision.

Without information about the consequences of the uniformity provision, we think that making a socially wise midcourse correction is difficult. Data are not available in the NCWQ findings about policy alternatives to the 1983 goal other than the recommendation to abandon or delay it in favor of the 1977 goal. In reality there are several alternatives to the 1983 goal. This book analyzes the consequences of several nonuniform alternatives to the 1983 goal and recommends a potential midcourse correction based on a limited analysis of the many facets of the 1972 Law.

The findings and recommendations in this book are seen as a necessary complement to rather than a substitute for the work of the NCWQ. This book could not begin to address the range of issues and insights presented in the NCWQ report. The book is only an attempt to ensure that sufficient data are available to evaluate the consequences of the 1972 Law. In addition, it is an argument for the necessity of a systematic data base for evaluating the consequences of any national water quality management policy.

To achieve compatibility with NCWQ studies and other sources, where possible all NRDI cost estimates are reported in 1975 dollars. Exceptions are individually noted.

REFERENCES

Library of Congress. 1973. A Legislative History of the Water
    Pollution Control Act Amendments of 1972. Washington, D.C.:
    Government Printing Office.
Luken, Ralph A., Daniel J. Basta, and Edward H. Pechan. 1976.
    The National Residuals Discharge Inventory. Washington,
    D.C.: National Research Council.
Public Law 92-500. 1972. 86 Stat. 816-904. Washington, D.C.:
    Government Printing Office.

2

NATIONAL RESIDUALS
DISCHARGE INVENTORY

The NRDI is a systematic procedure for evaluating various aspects of the 1972 Law. The inventory is structured to predict potential reductions in residuals discharged into the environment and the associated costs of the application of uniform technologies as defined by EPA for municipal and industrial sources of residual. It can compare the resulting reductions in discharges from these sources with those from other sources, primarily urban stormwater runoff and nonirrigated agriculture, by river basin (defined as either 99 river basins or 18 regions comprising the continental United States). More importantly, NRDI allows for an evaluation of policy alternatives to the uniform application of control technologies to point sources as defined in the act. These policies reflect alternatives where, in a given river basin, achievement of the 1983 effluent limitations would not make a significant improvement in total residuals reductions and water quality, and where a given level of residuals reduction could be achieved at a lower cost without the uniform application of control technology to point sources.

In this chapter, the major components of the NRDI are introduced and the major design and generic limitations inherent in the NRDI are discussed.

NRDI COMPONENTS

NRDI consists of (1) inventories of production (and consumption) activities that discharge residuals, (2) a system for estimating increased industrial production and population growth, (3) water pollution abatement policies that include the BPT/ST and BAT/BPWTT

FIGURE 2.1

NRDI Components

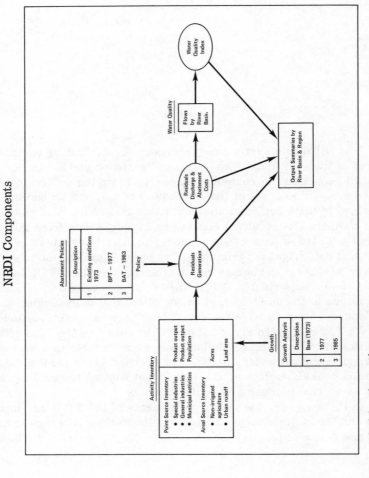

goals of the act, and (4) a model for computing a water quality index. Figure 2.1 relates each of these components; each of the blocks in the diagram is briefly described below and documented in more detail in the following chapters.

Activity Inventories

The purposes of the activity inventories are to relate process/ production data to residuals generation coefficients and to calculate residuals generation for a particular activity. The specificity of the individual inventories for the above process depends upon the importance of each sector as a residuals generator and the availability of data.

There are activity inventories for municipal, industrial, and areal categories. The municipal category includes a sewage treatment plant inventory. The industrial category includes an in-depth industry inventory for the significant process water users and a general industry inventory for the vast majority of other sources of industrial residuals. The areal category includes an urban runoff inventory for 253 standard metropolitan statistical areas (SMSA's) and an inventory of all nonirrigated cropland in the contiguous United States. Where appropriate and available, information about these activities includes location of activity, measures of production (physical output, employees, land area, or population), type of production process, product output, and current abatement technologies in place. The primary emphasis in the development of NRDI has been on estimation of residuals generation (gross residuals), residuals discharge (net residuals), and the costs of abatement technology under base case (1973) conditions. However, estimates of residuals and costs for a single year are not sufficient to evaluate some of the significant issues mentioned in Chapter 1. Consequently, an effort has been made to integrate growth projections into NRDI.

Growth Analysis

The spatial pattern, types, and intensity of man's activities change over time, and the various NRDI activity inventories are formulated to account for some of these impacts on residual generation.

In the NRDI, agricultural and urban runoff do not grow. Agricultural residual generation does not change, because there is no way in the NRDI of estimating the type of land or the cultivation

techniques that might be used in the future. While there is a con-
sensus among experts that agricultural production will increase
in the future, the relationship between incremental production and
incremental residual generation and discharge is not known. For
this reason, the NRDI scenarios for the future assume static residual
generation for agriculture. For urban stormwater, the other static
activity, residual generation is expected to change only slightly,
because the base case includes residuals from both sewered and
unsewered areas in urban counties. The combined sewer overflow
portion is also not expected to grow, because no combined sewers
are being built.

For industry, the projected growth is based on increases in
physical output for manufacturing activities by major industry group.
The growth analysis design permits either national or regional
averages for growth; in this discussion, national industrial growth
is used. The projected growth for municipalities is based on population growth reported for each treatment facility.

Information on projected growth for industry is available
either from the Wharton analysis used by NCWQ (Wharton Econometric Forecasting Associates 1975) or the U.S. Departments of
Agriculture and Commerce, Office of Business Economics and
Economic Research Service (OBERS) series E (Water Resources
Council 1974a). The projected growth for municipalities is based
on U.S. Department of Commerce, census series E population
growth rates (Water Resources Council 1974b). Other factors not
included but which may be significant, are information on changes
in process water use over time, in technology, or in product output
specification.

Alternative Policies

Alternative abatement policies are the key inputs that drive
the NRDI to a solution. The initial policies evaluated are the
application of BPT/ST and BAT/BPWTT objectives. The BPT and
BAT objectives for industrial activities are the EPA-defined series
of technological processes for each industrial subcategory. The ST
and BPWTT objectives for municipal activities are the series of
technologies requested in the 1974 Needs Survey (Environmental
Protection Agency 1975) on secondary treatment and on processes
more stringent than secondary treatment, respectively. There are
no EPA-defined technological objectives for either urban runoff or
nonirrigated agriculture.

Each alternative policy is specified as a set of unit processes for reducing gross residuals to net residuals. Each unit process has a known removal efficiency (percent reduction) and cost. The cost may be identified as a capital or initial investment cost, operations or maintenance cost, or as an annual cost. This book reports only the capital cost.

Several policies other than BPT/ST and BAT/BPWTT can be evaluated by the NRDI. One policy assumes technological objectives for urban runoff and nonirrigated agriculture. The future technological objective for urban runoff is based on residual reduction technologies identified by an NCWQ contractor. The technological objective for nonirrigated agriculture is defined as a set of conservation practices identified by the U.S. Department of Agriculture. A second policy exempts point-source activities from either BPT/ST or BAT/BPWTT requirements if these activities are estimated to be ocean dischargers. A third policy exempts point-source activities from BAT requirements if these activities discharge into river basins with sufficient assimilative capacity. A fourth policy is a minimum-cost solution (limited to changes in residual reduction technology) for meeting specified biochemical oxygen demand (BOD) discharge limits and concentrations.

Water Quality Index

The purpose of the water quality indexing procedure is to convert the information on net residuals into one measure of water quality-BOD concentrations. The procedure involves computing the concentration resulting from net BOD residuals. The computational procedure ranks basins according to concentration. Such a ranking may then be used to identify those basins in which water quality is relatively better or worse under current conditions and which may be significantly affected by different water quality management policies. Since the average conditions do not reflect a real situation at any given location or time, they cannot be used to attempt to pinpoint specific water quality problems in a subbasin or stream segment. Water quality-related data unique to each river basin are approximations based on low and average flow conditions.

The outputs resulting from the interaction of the four NRDI components are gross residuals, net residuals, abatement costs, and water quality index.

These outputs must be taken together to evaluate a given policy, because no one output is by itself an adequate evaluation measure. However, even this combination is no substitute for basin-specific evaluation. At that level of detail, data are available for assessing actual changes in physical, chemical, and biological parameters and for measuring the damages that are sustained directly by human beings or indirectly by plants and animals of value to man.

Previous Supporting Studies

The tasks of constructing, validating, and operating the NRDI as well as preparation of the National Research Council final report on the NRDI were completed in about nine months. This is a rather short period in which to perform all the tasks needed to complete the analysis. Fortunately, existing models were available to serve as patterns for the industrial and municipal activity inventories.

Both the industrial and municipal activity inventories are conceptually based on work performed by the authors at EPA in preparing the Economics of Clean Water (Environmental Protection Agency 1973). These analyses focused on computing the costs of meeting BPT/ST requirements, using techniques to estimate flow and then capital costs. Since these techniques had been used before and validated, they were used in the NRDI. The major supplement was to add residuals information to the analyses, a task accomplished by adding information on residual concentrations before treatment and resulting efficiency, that is, removal rate, for each of the unit treatment processes. Similarly, the NRDI water quality index was conceptually based on the study by Nathaniel Wollman and Gilbert Bonem (1971), although the approach adopted was somewhat simpler.

While the above two studies had the strongest influence on the structure of the NRDI, a large number of other studies influenced our approach to some degree.

Geographic Units of NRDI

NRDI contains a quantitative assessment of residuals generation, discharge, and costs in each of the 3,111 counties or county approximations in the contiguous United States. Data in the industrial, municipal, urban runoff, and nonirrigated agriculture inventories are available for each county, if the activity is present in that county. However, the data are not displayed at the county level, but rather are aggregated for purposes of analysis for the 99 river basins and 18 regions, and for the nation. River basin and region boundaries are hydrologic units as approximated by county boundaries.

The river basins used in this book are the aggregated subareas as defined by the Water Resources Council (1970). The boundaries of the river basins representing the contiguous United States are illustrated in Figure 2.2 and their names and characteristics are listed in Appendix B.

The regions used in this book are the water resource regions defined by the Water Resources Council. One or more river basins make up each region. The boundaries of the 18 regions in the contiguous United States are also illustrated in Figure 2.2 and the names are also in Appendix B.

CONCEPTUAL BASIS

In developing the NRDI, the authors made an effort to avoid some of the pitfalls of large-scale modeling. Douglass Lee (1973), in his review of the limitations of large-scale models, concluded that the most important attribute of any useful model should be transparency. The model should be not the usual "black box," but readily understandable after a reasonable investment of effort. Admittedly, a model can be transparent and wrong at the same time, but at least the interested parties can determine why they disagree with its conclusions.

To achieve this transparency, NRDI was built upon the guidelines enumerated by Lee:

1. NRDI is designed to address a selected set of issues in P. L. 92-500, primarily the cost and effects of different effluent limitations on all river basins in the United States. The issues were clearly defined and then a methodology was developed for addressing them rather than a methodology for finding a set of issues. While this decision limits the number of issues that the NRDI can address, it does mean that the NRDI can more adequately address these particular issues.

2. NRDI is primarily an empirical construction recognizing the complexity of residuals generation and the potential range of impacts. Its use can provide insight into questions of uniformity and diversity. NRDI uses as much data as are available, but it does not attempt to address an issue in a manner more sophisticated than the state of the data and modeling capacity would allow.

3. NRDI is a simple model, as will be evident in the following chapters. It uses only those functional relationships needed to measure gross and net residuals, the costs of technologies, and impacts on water quality. It abstracts from reality to the degree that it does

FIGURE 2.2

River Basins and Regions

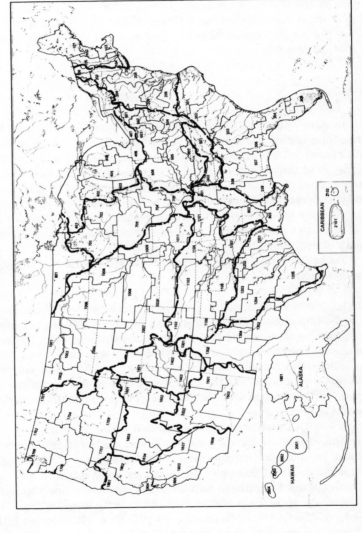

Source: Water Resources Council. 1975. Washington, D.C.: Water Resources Council.

not take into account all possible variations in gross residuals, the increased residuals discharges resulting from specific interindustry linkages in the economy, the water quality impacts of residuals discharged into a given river segment, or intermedia residuals relationships.

DESIGN AND GENERIC LIMITATIONS

The NRDI was designed to simulate some of the effects resulting from implementation of national water pollution control policies. While the system meets its specific objectives, its more general applicability to other water pollution issues is constrained in two ways; first, by conscious decision in the design stage to limit the scope (and complexity) of the system to issues that could be,meaningfully and accurately quantified; and second, by inherent limitations in the range and accuracy of currently available information needed to analyze issues of possible interest. The discussion in this section includes only those limitations that are applicable to the NRDI as a whole. More specific limitations are discussed in later chapters.

NRDI is not structured as a direct optimization procedure. Rather, it is designed to assess quickly the impacts of alternative specified national policies, as these affect the various sources of residuals. NRDI does this by computing the quantities and costs of reduction for each discharger or group of dischargers, and the aggregate effect on water quality in each river basin of the resulting discharges. NRDI does not account for the modification and discharge of secondary residuals, except for the costs of dewatering and disposing of sludges (landfill).

Thus, the system does not account for environmental or economic implications related to the ultimate disposal of residuals to other media, nor does it account for costs of meeting water pollution standards for residuals originating in other media (for example, scrubber sludges).

Finally, the water quality model was designed specifically for national policy evaluation and thus is not as sophisticated as other more detailed models.

Residuals other than BOD and suspended solids (SS) are not considered in the NRDI due to the lack of adequate information both on sources and effectiveness of control processes. * For

---

*The exceptions are nitrogen (N) and phosphorus (P) studied for municipal and nonirrigated agriculture sources.

example, this limitation means that generation of toxic residuals as well as effects and costs of control practices required, are not considered. This limitation could be significant both from a water quality and an economic perspective.

Economic and environmental effects of more stringent treatment requirements for industry (required by water quality standards) and of controls for areal sources of residuals (required by areawide planning) are not included in the analysis due to the complete absence of a supporting information base.

Damages associated with current levels of water pollution as well as benefits from changes in abatement levels cannot be estimated due to considerable uncertainty concerning relationships between effluent limits for selected residuals, water quality changes in those residuals, and changes in beneficial use as a function of the changes in ambient levels of selected residuals.

While the design limitations discussed above were built into the NRDI under the belief that their value in terms of additional insights did not justify the additional cost and complexity, the generic limitations had to be accepted whether or not their inclusion would significantly change the analysis results. The existence of such key gaps in our knowledge supports the need for research programs structured to answer remaining information questions, and establishment and implementation of regulatory policies that are flexible enough to change as more information becomes available.

REFERENCES

Environmental Protection Agency. 1973. Report to the Congress: The Economics of Clean Water. Washington, D.C.: Environmental Protection Agency.

_____. 1975. Report to the Congress: Cost Estimates for Construction of Publicly Owned Wastewater Treatment Facilities—1974 Needs Survey. Washington, D.C.: Environmental Protection Agency.

Lee, Douglas. 1973. "Requiem for Large-Scale Models." American Institute of Planners Journal, May.

Water Resources Council. 1970. Water Resource Regions and Subregions for the National Assessment of Water and Related Land Resources. Washington, D.C.: Water Resources Council.

_____. 1974a. 1972 OBERS Projections. Washington, D.C.: Government Printing Office.

_____. 1974b. Series E OBERS Projections and Historical Data: Population, Personal Income, and Earnings—Aggregated Subareas. Washington, D.C.: Water Resources Council.

_____. 1975. Map of Aggregated Subareas and Water Resource Regions. Washington, D.C.: Water Resources Council.
Wharton Econometric Forecasting Associates. 1975. Wharton Econometric Forecasting Estimate, Mark IV. Philadelphia.: Wharton Econometric Forecasting Associates.
Wollman, Nathaniel, and Gilbert Bonem. 1971. Outlook for Water— Quality, Quantity, and National Growth. Baltimore: Johns Hopkins Press.

CHAPTER
3
INDUSTRIAL
ACTIVITIES

Industrial activities are analyzed by either an in-depth or a general approach to an individual industry. For an in-depth study, an industry's inventory is uniquely defined by predominant production process, product output in base year, and in-place treatment facilities. Residuals generation coefficients are keyed to type of production process. For a general study, the inventory uses data only on the number of employees per plant, the number of plants by employment range per county, a uniform estimate of in-place treatment facilities, and a national estimate of residual generation. For both in-depth and general studies, residuals discharge and control costs are estimated for plants that only discharge directly into surface waters and not for plants that discharge indirectly through municipal treatment plants.

Industries that account for the majority of industrial residuals discharge and associated control costs are studied in depth. All other industries are studied in a less rigorous general manner. It was assumed that only a minimal increase in the accuracy of the national and regional assessments would result from studying these industries in more depth.

INDUSTRY STUDY CRITERIA

The major criterion for determining whether an industry should be studied in depth is significance of residuals generation. However, since there is no systematic data collection of actual residuals generation, process water intake (the water utilized directly in the production process, as opposed to cooling water) was used as a proxy.

Bureau of the Census data (U.S. Department of Commerce 1971) on process water intake show that a very limited number of industrial categories (measured by the four-digit Standard Industrial Classification or SIC code) and of plants account for a significant amount of process water intake. The top six four-digit SIC categories constitute 10 percent of the plants with water intake of over 20 million gallons per year and 67 percent of the process water intake. The top 15 four-digit SIC categories constitute 21.2 percent of the plants and 77 percent of process water intake. To include an additional 7 percent of process water intake, the number of SIC categories to be analyzed would have to double.

Given the data requirements to perform an in-depth analysis, another selection criterion is data availability. NCWQ technology contractor reports, EPA files, and industry directories were used to identify specific plant locations, types of production processes, and plant capacities. Although NCWQ technology contractors collected sufficient data on most of the industries they studied, in some cases the data were either coded to avoid identifying plant location or were deemed inadequate.

Based on these criteria, eight industries were selected for in-depth analysis (including the top six four-digit SIC codes). These sources account for approximately 74 percent of total process water intake.

The remainder of manufacturing industries, representing the majority of plants but only 26 percent of total process water intake, are studied less rigorously as general industries. Table 3.1 shows the selection of industries for either in-depth or general NRDI study, using the NCWQ format, and shows the other categories that have been identified as water-using industries.

The analysis described here, like any other analysis, includes numerous implicit and explicit assumptions. Key assumptions applicable to both in-depth and general analysis as well as those applicable to only one type of analysis or the other are given in Table 3.2.

METHODOLOGY FOR IN-DEPTH STUDIES

The in-depth industry studies consist of separate analyses of each of the eight significant industries identified in Table 3.1. This section explains the general analysis methodology and shows key characteristics of each industry analysis. Estimated residuals generation and discharge, replacement value of in-place facilities, and the cost of BPT and BAT technologies for each industry are given in tabular form later in the chapter.

TABLE 3.1

Industries Selected for In-depth or General Study

Industries Selected

In-depth Study

Pulp and paper[a b]	Plastics and synthetics[a b]
Petroleum refining[a b]	Organic chemicals[a b]
Textiles[a b]	Inorganic chemicals[a b]
Iron and steel[a b]	Steam/electric power[a b]

General Study

Ore mining and dressing[b]	Leather[b]
Coal mining[b]	Glass[b]
Petroleum and gas extraction[b]	Cement[b]
Mineral mining and processing[b]	Structural clay
Fish hatcheries	Pottery[b]
Meat products and rendering[b]	Concrete, gypsum[b]
Dairy products[b]	Asbestos[b]
Grain mills[b]	Fiberglass
Cane sugar processing[b]	Ferroalloy[b]
Beet sugar processing[b]	Nonferrous metals[b]
Seafood[b]	Transportation
Timber products	Water supply
Furniture and fixtures	Steam supply
Builders' paper[b]	Auto and other laundries
Paint and ink	Foundries
Soap and detergent	Nonferrous mill products
Phosphates	Miscellaneous food and beverages
Fertilizer[b]	Machinery
Paving and roofing[b]	Electroplating[a b]
Rubber[b]	Fruits and vegetables[a b]
	Miscellaneous chemicals[a b]

---

[a]Studied in depth by NCWQ.
[b]Included in NRDI analysis.
Sources: National Research Council. 1975. National Residual Discharge Inventory Computer Outputs. (hereafter referred to as NRDI Computer Outputs) Washington, D.C.: National Research Council; and National Commission on Water Quality. 1976. Staff Report to the National Commission on Water Quality. Washington, D.C.: Government Printing Office.

TABLE 3.2

Key Industry Analysis Assumptions

Assumptions
Applicable to In-depth and General Analysis Pretreatment costs for municipal dischargers are not computed. Pretreatment adjustment is only applied to industries with 25 percent or more of process wastewater discharged to municipalities. Only end-of-pipe treatment options are considered. EPA-specified BPT and BAT unit processes meet EPA effluent guidelines. Regional or stochastic differences in residuals generation and treatment are minor and can be ignored. Treatment processes operate at design efficiency. All discharge is treated by each unit process.
Applicable Only to In-depth Analysis All plants operate at or near full capacity. Residuals generation coefficients are a function of predominate process and not affected by plant size. Plant age is not significant.
Applicable Only to General Analysis Residuals generation is proportional to manufacturing employment. Differences are insignificant within a four-digit SIC category for production process, product specifications, and age of equipment. All plants in a four-digit SIC category have the same treatment in place. Plants with less than 20 employees are omitted as being insignificant.

Source: Pechan, E. H. 1976. Program Documentation: The National Residuals Discharge Inventory. Washington, D.C.: National Research Council.

The general methodology, shown in Figure 3.1, can be briefly explained in four steps.

The first step is the development of the most complete inventory available of the plants existing in each industry in 1973 and their subcategorization by predominant production process. The inventory includes data on product output per plant, usually measured in tons or output units per day, and actual or estimated in-place treatment technologies per plant. If 25 percent or more of wastewater discharge in the industry is discharged exclusively

FIGURE 3.1
Methodology for In-depth Industry Analysis

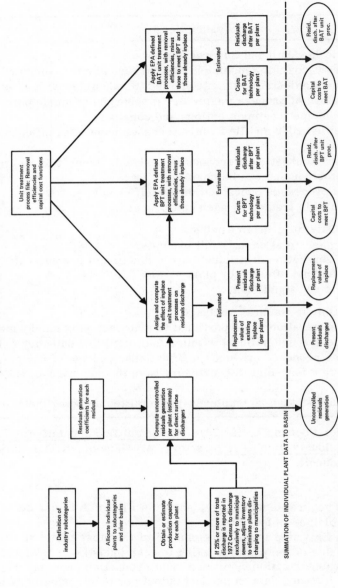

Source: Pechan, E. H. 1976. Program Documentation: The National Residuals Discharge Inventory.
Washington, D. C.: National Research Council.

24

TABLE 3.3

Characteristics of In-depth Industries

Industry	Number of Subcategories	In-Place Treatment Specification			BPT and BAT Specification		Adjustment for Municipal Discharge	Plant with More Than One Product
		By Plant	By Industry Subcategory	By Industry	By Industry Subcategory	By Industry		
Pulp and paper	13	X				X		
Organic chemicals[a]	64			X		X		X
Petroleum refining[b]	5	X				X		
Iron and steel	5		X		X			
Inorganic chemicals	35			X	X			X
Plastics and synthetics	17			X		X		X
Textile mills[c]	11		X		X		X	
Steam/electric power	3		none in place			X		

[a]Costs and residuals adjusted by scaling production at each plant by 4.25.
[b]For this industry, 5 percent of flow is treated by "petroleum flotation," 50 percent by cooling tower and evaporation.
[c]Employee data were used to estimate flow for 1,926 plants based on regressions for 382 plants with product data; municipal discharge estimated by state.

<u>Source</u>: Compiled by the authors.

to municipalities, the inventory is adjusted to eliminate plants
discharging exclusively to municipalities.

Otherwise, all plants in the industry are assumed to be direct
surface dischargers. The information on in-place treatment technologies, combinations of specific unit processes, are taken from
NCWQ and EPA development documents. Where information about
specific plants is unavailable, the average reduction technology
estimated to be in place in a subcategory is assumed to apply to all
plants in the subcategory.

The second step is to assign residual generation coefficients,
developed by production process and based upon product output, to
each subcategory and compute residual generation. The residuals
estimated are BOD, SS, and wastewater flow. The coefficients are
measured in pounds per ton of output or per ton of raw material
processed, for process water, which is measured in gallons per
ton of output. The residual generation coefficients are based on
either NCWQ technology reports or EPA development documents
for a particular industry.

The third step is to compute residual generation and the
residual discharges in 1973 and after BPT and BAT for each plant;
and the value of residual reduction technologies in place in 1973 and
the cost of BPT and BAT technologies. The set of unit treatment
processes used, and their associated removal efficiencies and costs,
are shown in Appendix C.

The fourth step is to summarize the individual plant data for
whatever level of aggregation desired: geographic (county, basin,
SMSA, region) and/or level of industry detail (subcategory, group
of categories.)

Although each of the industries is fundamentally analyzed
following the same general methodology shown in Figure 3.1, they
all exhibit slightly different characteristics as shown in Table 3.3.

METHODOLOGY FOR GENERAL STUDIES

All industries other than those studied in depth are analyzed
as described in this section and are grouped under the 41 categories shown in Table 3.1. Four-digit SIC categories are used as
the basic unit of analysis. The general methodology used is very
similar to the in-depth study method, once residual generation per
plant has been estimated.

The 41 industry categories analyzed are identified by 411
four-digit SIC categories. Table 3.4 shows the four-digit SICs
identified in each category and which SICs are included in the NRDI

TABLE 3.4

Four-Digit 1972 SIC Categories Associated
with General Industry Categories in NRDI Analysis

General Industry Categories	Four-Digit SIC Categories
Ore mining and dressing	1011, 1021, 1031, 1094, 1041, 1044, 1051, 1061, 1092, 1099
Coal mining	1111, 1112,[f] 1211, +1213
Petroleum and gas extraction	1311[a], 1321,[a] 1381,[a] 1382,[a] 1393[a]
Mineral mining and processing	1411, 1422,[e] 1423,[e] 1429,[e] 1422, 1446, 1452,[e] 1453,[e] 1454,[e] 1455, 1496, 1459, 1499, 1479,[e] 1472, 1473, 1474,[e] 1475, 1476, 1477, 1492
Meat products and rendering	2016, 2017,[b] 2011, 2013, 2077[c]
Dairy products	2021, 2022, 2023, 2024, 2026
Grain mills	2046, 2043, 2041, 2044, 2045, 2047,[e] 2048
Cane sugar processing	2061, 2062
Beet sugar processing	2063
Seafood	2091, 2092, 2077[c]
Builders' papers	2661[d]
Fertilizer	2873, 2874
Paving and roofing	2951, 1611,[e] 2952, 3996
Rubber	3011, 3021, 3031, 3069, 3293, 3822, 7534[e]
Leather tanning	3111
Glass	3211, 3221, 3229, 3231
Cement	3241
Pottery	3261, 3262, 3263, 3264, 3269[f]
Concrete, gypsum	3271,[f] 3272,[f] 3273,[f] 3275, 3274,[f] 3297
Asbestos	3272, 3293[f]
Ferroalloy	3313
Nonferrous metals	3331, 3332, 3333, 3334, 3341
Electroplating	3471, 3479
Fruits and vegetables	2032, 2033, 2034, 2035, 2037, 2038
Miscellaneous organic chemicals	2831, 2833, 2834, 2861, 2879, 2891, 2892, 2895, 2899

[a]Covered under two-digit SIC 13.

[b]2017 is also classified under miscellaneous food and beverages.

[c]2077 is classified under three industry categories including miscellaneous food and beverages.

[d]2661 is classified under two NCWQ industry categories.

[e]Not included in NRDI general industry inventory because of insufficient data.

[f]Not included in NRDI general inventory because not covered by NRDI unit treatment processes.

Note: Total number of four-digit SIC categories: 411; total number of four-digit SIC categories in NCWQ general industry categories: 394; total number of four-digit SIC categories included in machinery category (none of which is included in NRDI): 171.

Source: Compiled by the authors.

analysis. The NRDI analysis covers a portion or all of 24 categories encompassing 108 out of 143 four-digit SICs in these categories.

There were several reasons for the exclusion of certain industry categories, either partly or totally. Nine categories were not considered due either to lack of data, that is, residual generation concentrations, product output, or to the inability to estimate the cost, and effect on residuals discharges, of treatment technologies (in-plant, reuse, and recycle technologies) not included in the set of available unit treatment processes. The nine categories are soap and detergent, transportation, miscellaneous food and beverages, phosphates, timber products, furniture and fixtures, fiberglass, auto and other laundries, and paint and ink. Five additional categories were not investigated at all because the NCWQ estimated a low total cost for BPT and BAT technologies. They are steam supply, fish hatchery, nonferrous mill products, structural clay products, and foundries. Two categories, machinery and mechanical products, and water supply, were not investigated because of lack of data and the absence of final EPA guidelines for the industry. Deletion of these categories is unfortunate because the NCWQ has estimated high BPT and BAT costs for these industries. The consequences of not including machinery and mechanical products, and water supply in the NRDI data base are discussed later in this chapter.

The only plant-specific data involved in this analysis are employees per plant and plant location. All other information is estimated at the national level and distributed among plants by computing plant employee share of the national total within an SIC category.

The method can be briefly explained in several steps as shown in Figure 3.2. The first step, definition of industry categories or groups by four-digit SICs, has already been explained. The entire analysis is performed by SIC category and not until the very end is information by SIC aggregated for the appropriate industry category. The second step is to estimate total national residual generation for each SIC and distribute it to individual plants on the basis of employment. Generation is estimated based upon 1972 process water intake, which is used as an estimate of wastewater discharge and residual generation concentrations taken from several sources.

The third step is to estimate existing in-place, BPT, and BAT technologies as series of unit treatment processes that can reasonably be assumed to be uniformly applicable to all plants in an SIC category. Uniform in-place unit treatment process technologies for each SIC were primarily estimated using table 9

FIGURE 3.2

Methodology for General Industry Analysis

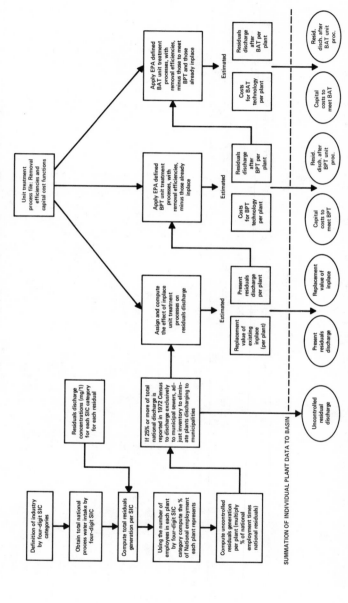

Source: Pechan, E. H. 1976. Program Documentation: The National Residuals Discharge Inventory. Washington, D.C.: National Research Council.

29

TABLE 3.5

1973 Residuals Generation and Discharge by Industry

Industry	Residuals Generation in 1973 (million pounds)		Residuals Discharge in 1973 after applying In-Place Control Technologies (million pounds)		Residuals Discharge in 1973 after Applying BPT (million pounds)		Residuals Discharge in 1973 after Applying BAT (million pounds)	
	BOD	SS	BOD	SS	BOD	SS	BOD	SS
Pulp and paper	3,696.6	4,876.6	1,898.5	2,234.3	369.7	390.1	184.8	146.3
Organic chemicals	1,840.1	NA	1,232.9	NA	173.0	NA	36.8	NA
Petroleum refining	211.8	62.8	113.6	29.1	10.6	1.4	4.2	0.5
Iron and steel	78.6	12,091.9	39.3	2,549.3	5.9	814.2	1.2	67.6
Inorganic chemicals	7.5	2,385.9	5.0	1,145.2	3.6	691.0	2.6	344.6
Plastics and synthetics	302.3	226.3	202.6	108.6	30.2	24.9	6.0	2.3
Textile mills	256.2	168.6	126.2	109.8	25.6	18.5	7.7	3.5
Steam/electric power	NA	5,877.0	NA	5,877.0	NA	816.9	NA	816.9
Total In-depth Industry*	6,400.0	25,700.0	3,600.0	12,000.0	620.0	2,800.0	240.0	1,400.0
Ore mining and dressing	0.0	1,090,555.0	0.0	688,863.7	0.0	8,724.5	0.0	8,724.5
Coal mining	0.0	127.3	0.0	40.4	0.0	5.1	0.0	5.1
Petroleum and gas extraction	10.7	15.2	10.7	15.7	6.4	3.0	6.4	3.0
Mineral mining and processing	0.0	30,745.2	0.0	21,104.4	0.0	885.1	0.0	884.3
Total Mining	10.7	1,121,438.0	10.7	710,021.9	6.4	9,617.6	6.4	9,616.9
Meat products and rendering	250.3	187.9	30.3	8.0	25.2	6.3	12.6	1.8

Dairy products	41.1	16.7	41.1	16.7	10.3	5.8	5.1	1.6
Grain mills	104.2	52.7	19.1	5.0	17.1	4.1	8.6	1.2
Cane sugar processing	15.3	164.4	11.1	83.7	6.6	46.5	0.4	4.1
Beet sugar processing	39.9	139.2	9.2	48.7	6.2	23.4	2.5	7.5
Seafood	24.5	19.5	24.5	19.5	6.1	6.8	3.7	1.4
Builders' paper	19.2	65.3	19.2	65.3	3.2	11.0	1.6	3.1
Fertilizer	0.0	0.0	0.0	0.0	0.0	0.0	0.0	0.0
Paving and roofing	0.0	5.9	0.0	2.8	0.0	1.8	0.0	1.9
Rubber	22.1	70.1	22.1	70.1	2.8	7.4	1.4	4.3
Leather tanning	60.4	89.7	40.5	43.0	10.1	15.1	5.1	4.2
Glass	0.7	3.1	0.6	2.6	0.6	2.6	0.4	1.4
Cement	103.7	94.2	103.7	18.8	69.5	9.0	69.5	9.0
Pottery	0.0	30.5	0.0	30.5	0.0	8.9	0.0	2.8
Concrete, gypsum	0.1	1.5	0.0	1.3	0.0	0.6	0.0	0.6
Asbestos	0.0	11.0	0.0	5.3	0.0	3.2	0.0	3.0
Ferroalloy	0.0	2.2	0.0	1.0	0.0	0.6	0.0	0.2
Nonferrous metals	0.0	16.3	0.0	8.7	0.0	5.0	0.0	5.0
Electroplating	0.3	1.6	0.3	1.6	0.1	0.1	0.0	0.0
Fruits and vegetables	410.7	255.6	275.2	122.7	58.8	42.6	24.4	11.9
Miscellaneous chemicals	729.7	368.8	289.7	150.5	73.3	40.7	22.8	8.8
Total Other Industry	1,819.1	1,596.1	886.5	705.9	299.9	241.6	168.0	74.8
Total General Industry*	1,800.0	1,123,000.0	900.0	711,000.0	310.0	9,900.0	170.0	9,700.0
Total All Industry*	8,200.0	1,149,000.0	4,500.0	723,000.0	930.0	12,700.0	410.0	11,100.0

*Rounded.

Note: NA means data not available (and not significant).

Source: NRDI Computer Outputs. 1975.

31

in the 1972 Census of Water Use in Manufacturing (U.S. Department
of Commerce 1975). In some cases, EPA development documents
were used. BPT and BAT technology assumptions were taken exclusively from EPA development documents. The remaining steps
for computing control costs and related residual discharges per
plant and aggregating these data at the appropriate levels are the
same as for the in-depth studies. The same set of unit process cost
functions as used for the in-depth studies are used here.

BOD and SS residuals generated in 1973 as well as residuals
discharged in 1973 after application of in-place, BPT, and BAT
technologies are shown in Table 3.5. Costs associated with implementing these technologies are displayed in Table 3.6.

GROWTH IN INDUSTRIAL RESIDUALS GENERATION

The NRDI growth estimate for industries in the in-depth
and general studies is based on two-digit SIC growth projections.
Different national growth projections were available for use in the
NRDI from the Wharton Econometric Forecasting Associates and
from OBERS projections. Growth projections by basin are also
available from OBERS.

The Wharton national projections are based on a large-scale
macroeconometric model with an imbedded input/output table
(Preston and O'Brien 1975). This combination permits consideration of demand and supply simultaneously in long-term projections.

A final demand model provides estimates of constant-dollar
gross national product (GNP) and its disaggregated components,
including consumption, investment, exports, imports, and government spending. Key inputs to the consumption functions include
income, price, wealth, credit conditions, and tastes. The basic
data in the investment sector are obtained from the investment
surveys of Department of Commerce and the Securities and Ex-
change Commission. Government spending, including federal,
state, and local sectors, are derived from survey data.

A supply model provides estimates of physical output by
sector. Production functions, which are available for most sectors,
relate inputs, primarily manpower availability, to physical output.
The sectors include agriculture, mining, manufacturing (disaggregated by two-digit SIC), transportation, communications,
regulated industries, and commercial activities. The government
sector output is exogenuously determined.

An input/output model links the final demand and supply
estimates in order to estimate gross sales levels by industrial

TABLE 3.6

Technology Costs for Industry
(millions of dollars)

Industry	Replacement Value of Existing Facilities (1973)	Incremental Capital Costs of BPT	Incremental Capital Costs of BAT
Pulp and paper	1,996	1,793	633
Organic chemicals	339	1,341	1,302
Petroleum refining	316	650	1,241
Iron and steel	849	1,883	1,300
Inorganic chemicals	348	544	139
Plastics and synthetics	410	270	176
Textile mills	254	600	322
Steam/electric power	0	1,727	0
Total In-depth Industry*	4,100	8,800	5,000
Ore mining and dressing	18	74	0
Coal mining	5	39	0
Petroleum and gas extraction	0	138	0
Mineral mining and processing	16	29	4
Total Mining	39	280	4
Meat products and rendering	512	74	160
Dairy products	0	52	93
Grain mills	46	5	22
Cane sugar processing	8	36	40
Beet sugar processing	26	29	26
Seafood	0	64	41
Builders' paper	0	52	16
Fertilizer	61	118	94
Paving and roofing	76	4	2
Rubber	0	240	86
Leather tanning	23	89	16
Glass	38	12	75
Cement	1	68	0
Pottery	0	29	20
Concrete, gypsum	2	32	0
Asbestos	16	3	3
Ferroalloy	9	35	7
Nonferrous metals	67	35	26
Electroplating	4	260	65
Fruits and vegetables	223	117	200
Miscellaneous organic chemicals	803	1,323	751
Total Other industry	1,915	2,677	1,743
Total General Industry*	2,000	3,000	1,800
Total All Industry*	6,100	11,700	6,800

*Rounded.

Source: NRDI Computer Outputs. 1975.

sector to sustain final demand. The input/output model explicitly
includes a mechanism that uses information on relative sector
prices.

The Wharton GNP and gross product originating data for
1973 and projected data for 1980 and 1985 are displayed in Table
3.7 in 1967 dollars.

The OBERS national projections are calculated mainly from
the supply portion of the economy (the major exception is the
determination of requirements for food and fiber) (Water Resources
Council 1974a). The (GNP) projected for the nation is based on
population, labor force participation, military strength, hours
worked per man-year, and GNP per man-year. Projected national
population (Census Bureau Series E data) is the most significant
variable in the total projection process.

While GNP is the most comprehensive and widely used measure of the national economy, the OBERS data disaggregated by
industrial activity (SIC) and regions are available only as personal
income. The choice of using GNP or personal income rested on
three considerations. First, personal income has a close and
comparatively constant relationship to GNP; second, its regional
location is clear; third, it could be measured from available data
sources, and the methodology for preparing local area estimates
of personal income had already been prepared. The OBERS GNP
and personal income data for 1973, 1980, and 1985 are displayed
in Table 3.7.

The OBERS regional projections are the projected national
totals distributed in accordance with projected trends in the
regional distributions of economic activities. The distribution is
on the basis of four models: basic industries, except agriculture
and armed forces; agriculture; residentiary or service industries;
and population. The basic or export industry model was derived
from the "shift-share" technique of regional industrial analysis.
The agricultural model is based on an extension of trends of production from an historical base of 1947 to 1970. The residentiary
model is a multiplier effect, which relates service employment
or earning to basic employment or earnings. The population model
derives area population from area employment. All these models
are explained in detail by the Water Resources Council (1974a).

The OBERS regional personal income data for 1975, 1980, and
1985 are too numerous to reproduce in this study (Water Resources
Council 1974b). OBERS measured regional personal income data
for industrial categories for 1971 or 1973 are not available because
of disclosure restrictions.

TABLE 3.7

OBERS and Wharton Industrial Projections
(millions of 1967 dollars)

Industrial Groupings	Wharton Output			OBERS Output			OBERS Personal Income		
	1973	1980	1985	1973	1980	1985	1973	1980	1985
Mining	17.4	13.9	20.3	17.4	19.7	21.2	6.4	6.5	6.9
Metal (10)	1.7	1.0	1.2	1.7	1.8	1.9	0.9	1.0	1.0
Coal (11, 12)	2.3	2.9	4.2	2.3	2.4	2.5	1.8	1.8	2.0
Crude petroleum and natural gas (13)	11.8	8.5	12.4	11.8	13.7	14.8	2.6	2.5	2.6
Nonmetallic, except fuels (14)	1.6	1.5	2.4	1.6	1.8	2.0	1.1	1.2	1.3
Manufacturing	272.4	326.7	387.4	272.3	313.7	369.2	174.0	219.3	252.7
Food and kindred products (20)	23.9	28.3	28.0	23.9	25.1	27.6	13.3	16.0	17.4
Textile mill products (22)	10.2	12.0	14.0	10.2	9.7	11.0	6.0	6.7	7.4
Apparel and other fabric products (23)	8.8	9.1	8.4	8.8	9.8	11.3	6.8	8.7	9.8
Lumber products and furniture (24, 25)	9.8	10.6	12.5	9.8	11.0	12.7	8.1	8.9	10.0
Paper and allied products (26)	10.0	12.1	13.4	10.0	11.9	13.9	6.2	8.4	9.7
Printing and publishing (27)	10.7	14.3	16.1	10.7	14.5	17.2	9.3	13.0	15.3
Chemical and allied products (28)	27.0	32.3	36.8	27.0	33.6	41.9	10.8	15.6	18.8
Petroleum refining (29)	6.7	7.7	8.7	6.7	7.3	8.5	2.8	3.4	3.8
Primary metals (33)	17.4	21.2	29.2	17.4	16.4	17.1	14.1	14.3	15.3
Fabricated metals and ordnance (34, 19)	15.6	18.4	22.6	15.6	22.3	26.1	14.9	19.5	22.6
Machinery, excluding electrical (35)	27.7	32.1	42.2	27.7	29.3	33.7	20.3	24.5	28.1
Electrical machinery and supplies (36)	32.5	39.4	49.4	32.5	40.7	51.8	17.1	25.1	30.5
Motor vehicles and equipment (371)	26.0	29.5	33.2	26.0	28.6	33.6	14.1	15.5	18.0
Transportation equipment, excluding motor vehicle (37 except 371)	12.1	15.9	18.8	12.1	12.2	13.4	9.3	11.6	12.8
Other manufacturing (2, 30–82, 38, 39)	33.9	44.0	54.1	33.9	41.3	49.4	20.9	28.1	33.2
Gross national product	821.4	1,003.2	1,174.8	821.4	1,091.0	1,301.0	NA	NA	NA

Note: Numbers in parentheses are two-digit SIC codes. NA means data not available.

Sources: Wharton Economic Forecasting Associates. 1975. Wharton Econometric Forecasting Estimate, Mark IV. Philadelphia: Wharton Econometric Forecasting Associates; and Water Resources Council. 1974. 1972 OBERS Projections. Washington, D.C.: Government Printing Office.

35

At this point, a brief comparison of Wharton national projections and OBERS GNP and gross product originating projections will provide some insight into the uncertainty of projections for the period 1973 to 1985. Wharton projects a growth in GNP between 1973 and 1985 of 43 percent, or an annual rate of 3.3 percent. OBERS projects a growth in GNP for the same period of 55 percent, or an annual growth rate of 3.7 percent. (The OBERS 1985 projection is a linear extrapolation between 1980 and 1990, because there is no published estimate for 1985.)

The difference among the two-digit SIC categories is even more striking. For example, the Wharton estimate for primary metals is five times greater than the OBERS estimate; for petroleum, it is two times greater; and for textiles, it is one and one-half times greater. Because of these differences, either set of projections was usable in the NRDI.

Four industrial growth options are available in NRDI. In option 1, the Wharton national projections at the two-digit SIC level are used for each river basin. In option 2, the OBERS national projections at the two-digit SIC level are used for each river basin. In option 3, the OBERS/river basin projections are used at the two-digit SIC level. In option 4, the OBERS/river basin projections are modified to produce a national total equivalent to the Wharton projection; that is, the Wharton growth is allocated to river basins using the OBERS data.

Only option 1, the Wharton national projections at the two-digit SIC level, is used in the following chapters to estimate growth in residual generation from 1973 to 1985. The NRDI analysis is limited to one projection because of the difficulty of incorporating data from several projections into the text of the NRDI report, and because the different projections do not make a significant difference in the NRDI national estimates of future residual generation and costs. The NRDI analysis is based on the Wharton rather than the OBERS projections for two reasons. First, NRDI uses the same basic data as NCWQ wherever possible. Since NCWQ used the Wharton rather than the OBERS projection as their official estimate of industrial growth in the U.S. economy, NRDI incorporated the same data. Second, Wharton appears to give a more realistic projection than OBERS of U.S. industrial growth from 1973 to 1985. Given the economic situation in late 1975, an annual growth rate of 3.3 percent (Wharton) seems more plausible than 3.7 percent (OBERS).

Under all options, annual growth in each industry is assumed to be discharging residuals at the new source performance standards (NSPS) level, which is approximated by the application of

BAT. Thus, the growth projection for BPT combines the base-year estimate for discharge after application of BPT and the future-year estimate for discharge after application of BAT. The growth projection for BAT is strictly a scaler applied to the base-year discharge after application of BAT.

Capital costs for installing BPT and BAT to accommodate new growth are calculated on the basis of the increased residuals flow resulting from the above calculation. The procedure for calculating incremental cost may overestimate actual capital costs. Industrial growth in the future will probably use less water per unit output, if the past trend from the Census of Water Use in Manufacturing is an accurate estimate of future water use. In addition, industry may accommodate some new growth with the residual reduction technology installed to handle base-year production levels.

LIMITATIONS OF INDUSTRY STUDIES

The limitations of the NRDI industry studies may be classified into three categories: residuals estimation, cost estimation, and regionalization. There are limitations that apply to both in-depth and general study categories, to each specific study category, and to each individual industry.

A major limitation of NRDI industry studies is that only total wastewater, BOD, and SS are considered. Other perhaps more important residuals have not been included—chemical oxygen demand (COD), heavy metals, toxins. Although some data on other residuals for industry were available, the data were neither sufficiently accurate nor complete to permit analysis within the NRDI. As an example, the wide presence of heavy metal species in industrial waste discharges is illustrated in Table 3.8.

Estimations of total wastewater and BOD and SS residuals generation and discharges after application of control technologies are more accurate in the in-depth studies than in the general studies. Even so, the absolute accuracy of residuals estimation in the in-depth studies is difficult to attain. For example, to assume that all plants in an industry are operating at capacity tends to overestimate residuals generation. On the other hand, to assume that all treatment facilities operate with design efficiencies tends to underestimate discharge, a compensating effect.

Waste strength and volume data for industrial activities were based on average values from the literature. Variables that affect both strength and flow include process type, age and condition of equipment, and operating practice. Of these, only process type is considered explicitly in the model, by the subcategorization of the

TABLE 3.8

Heavy Metals Generation by Industry

Industry	Cadmium	Copper	Chromium	Lead	Mercury	Iron	Zinc	Nickel
Textiles			X					
Dyes		X	X					
Plasticizers			X					
Chlorine				X	X			
Titanium dioxide						X		
Sodium dichromate			X					
Rayon			X				X	
Latex		X	X				X	
Petroleum			X					
Tanning			X					
Iron and steel			X	X		X	X	X
Copper		X					X	
Lead	X			X	X		X	
Zinc	X						X	
Bauxite								
Aluminum		X						
Electroplating	X	X	X				X	X
Electric power		X	X		X		X	

<u>Sources</u>: EPA development documents.

special industries. Different process categories for the general
industries (four-digit SIC code) were not considered.

While the omission of age and operating practice as well as
other residuals-affecting factors reduces the utility of the generation and discharge information for plant-by-plant analyses, average
conditions should be relatively accurate over aggregations of as
many facilities as found in a river basin-sized area.

The difficulties of estimating generation are demonstrated in
Table 3.9; an NCWQ contractor and an EPA development document,
in considering organic chemicals, agreed on many of the BOD and
COD loadings, but disagreed significantly on several product processes.

While the special industry subcategorization is an attempt to
account for significant process/product-related differences in
residual generation, it cannot account for all of the waste variation. Table 3.10 shows the 10 percent to 90 percent probability
range for BOD effluent from an oil separator normalized to produce output for the five subcategories of the petroleum industry.
Note that even in cases where the medians are relatively close
(that is, for the petrochemical and integrated categories), the
ranges of strengths may be highly variable. Studies to explain this
type of variation have not been undertaken. The BOD variations for
the integrated category are applied in Table 3.11 to the discharge
projected by NRDI, which used the median value of 69.0. This
estimated a discharge of 113.6 million pounds per year, dropping
to 10.6 million at BPT and 4.2 million at BAT. The 10 percent
probability value for BOD (32.2 percent of the median value),
if applied to 1973 discharge, would estimate only 36.6 million
pounds per year. At the same removal rates, BOD discharge
would drop to 1.4 million pounds at BAT, which is only 1.2
percent of the 113.6 million pounds estimated for 1973 discharge
by NRDI.

In similar manner, the 90 percent probability value (3.1
times the median) produces a discharge of 354.0 million pounds
per year. Using NRDI removal rates, the discharge would drop
to 13.1 million pounds by applying BAT. So even if the 90 percent
value (which has only 10 percent chance of occurring) were the
actual value, BOD discharge at BAT would be 11.5 percent of the
1973 estimated discharge. This is a removal rate of 88.5 percent
compared to the 96.3 percent achieved using the median value.

Total SS show as much variation as BOD in this industry.
For the same integrated category, the SS values are: at 10
percent, discharge of 5.3 million pounds per 100 barrels of
feedstock; at median, discharge of 20.3 million pounds per 100

TABLE 3.9

Selected BOD and COD Waste Loadings
(pounds per 100 pounds of production)

Product/ Process	BOD		COD	
	EPA Estimate	Catalytic Estimate	EPA Estimate	Catalytic Estimate
Acetic acid	0.35	6.40	0.78	15.10
Acetone	0.26	1.00	1.10	4.00
Ethylene glycol	0.34	5.26	8.76	8.76
Ethylene oxide	0.70	3.80	6.48	7.70

Sources: Environmental Protection Agency. 1974a. Development Document for Major Organic Products. Washington, D.C.: Government Printing Office. and Catalytic Inc. 1975. Capabilities and Costs of Technology for the Inorganic Chemicals Industry to Achieve the Effluent Limitations of P.L. 92-500. Springfield, Va.: National Technical Information Service.

TABLE 3.10

BOD Effluent for Petroleum Wastes
(pounds per 100 barrels of feedstock)

Category	Probability		
	10 percent[a]	50 percent Median	90 percent[b]
Topping	0.45	1.2	76.0
Cracking	5.00	25.5	163.0
Petrochemical	14.30	60.0	715.0
Lube	22.00	76.0	265.0
Integrated	22.20	69.0	215.0

[a]Probability of this value or smaller = 10 percent.
[b]Probability of this value or smaller = 90 percent.
Source: EPA (1974b) Environmental Protection Agency. 1974b. Development Document for Petroleum Refining. Washington, D.C.: Government Printing Office.

TABLE 3.11

Effect of Variability in Residual Generation Coefficients—
Petroleum Refining

| | Discharge | | | Percentage of 1973 Median | |
Residual	1973 (million pounds)	BPT (million pounds)	BAT (million pounds)	BPT	BAT
BOD					
Median	113.6	10.6	4.2	9.3	3.7
10% (0.322)	36.6	3.4	1.4	3.0	1.2
90% (3.116)	354.0	34.3	13.1	30.2	11.5
SS					
Median	29.1	1.4	0.5	4.8	1.7
10% (0.261)	7.6	0.4	0.1	1.4	0.3
90% (3.892)	113.3	5.4	1.9	18.6	6.4

Sources: NRDI Computer Outputs. 1975; and Environmental Protection Agency. 1974b. Development Document for Petroleum Refining. Washington, D.C.: Government Printing Office.

barrels of feedstock; and at 90 percent, discharge of 79.0 million pounds per 100 barrels of feedstock. Using the median assumption, the 1973 discharge of 29.1 million pounds per year would drop to 0.5 million pounds at BAT, a removal rate of 98.3 percent. At the 10 percent probability value, the final BAT discharge would be 0.1 million pounds per year, while at 90 percent value it would be 1.9 million pounds per year. This last figure still represents only 6.5 percent of the 29.1 million pounds per year used in NRDI.

Even if the BOD and SS values were actually at the 90 percent probability-of-occurrence levels, the high rates of removal in the petroleum refining industry assure that the amount of BOD and SS discharged would be relatively minor compared to 1973 levels.

EFFECTS OF IN-PLANT CHANGES

The cost and residual analyses in the NRDI were based on imposition of end-of-pipe technology using historic wastewater and residual quantities. It is clear that as treatment requirements become more stringent, various in-plant changes to reduce flow and/

or residual loadings become more practical. Some in-plant changes would be possible only for new facilities while others can be implemented in existing plants.

In-plant changes that reduce flow and/or influent loadings would permit the construction of less costly end-of-pipe facilities. Since a plant is unlikely to implement changes unless the cost is below that of full end-of-pipe treatment, the estimates made by the NRDI can be considered as upper bounds.

While significant flow and/or residual reductions may be achievable in major industry categories and costs of meeting BPT and BAT limitations may be lowered, implementation costs of these changes as retrofits for existing plants may be significant relative to potential savings.

Whether an existing plant would make internal changes as opposed to simply adding end-of-pipe technology is subject to many factors including constraints on land for end-of-pipe technology, existing control technology in place, relative ease of implementation, plant shutdown required, and other advantages and disadvantages to making changes. It is far more likely that new facilities would build in these process changes and that the costs of doing so would be less than to retrofit.

In some cases, costs of flow and residuals reductions are justified for other savings. As an example, in the fruits and vegetables industry, a dry caustic peeling process has been shown to be both cost effective and advantageous in terms other than for water pollution control (EPA 1974c).

Thus, in-plant changes must be studied on an individual basis with consideration of many factors unrelated to water pollution control.

By definition, the general examination of industries is less precise than the in-depth examination. General industry residual data are less accurately estimated. NRDI estimates, as well as one desirable attribute of the NRDI general study approach is its consistency. Wastewater discharge data for each general industry are taken from the same source, 1972 Census of Manufacturers, and the same assumption is made for each: that process water intake is equal to wastewater discharge. It should be noted that using census wastewater data and typical discharge concentrations from EPA development documents produces estimates that tend to be low since plants using less than 20 million gallons per year are excluded from the census.

The NRDI regionalization of both industry residuals discharge and control cost data is believed to be the most accurate available to date. For the in-depth studies, the data are plant specific and the regionalization of residuals and cost is performed from the bottom up. The industries studied in depth represent the majority of total industrial residuals generation and control costs. On the other hand, for industries studied in general, national residuals estimates are disaggregated, based upon employment, for specific plants and then control costs are computed for each plant. The assumption that residuals generation is a function of employment has been used before in Economics of Clear Water (EPA 1973).

Limitations of NRDI Industry Cost Estimates

There are several important limitations of NRDI industry cost estimates. These limitations can be divided into two categories, those that tend to overestimate costs and those that tend to underestimate costs.

There are four important limitations that generally tend to cause overestimated costs. First, the NRDI estimates represent only end-of-the-pipe options; no estimates are made of in-plant changes. Second, total process wastewater discharge is, with one exception, run through all unit treatment processes in a treatment train. Third, most of the cost functions used were developed based upon the cost of wastewater unit treatment processes to be built for municipalities. Unit process costs are generally greater for constructing treatment systems for municipalities than for industries. Fourth, regional preferences in unit treatment process selection were not included. The use of less costly aerated stabilization basins, which were precluded in certain regions, instead of activated sludge systems would reduce costs. Despite these limitations, when compared to other sources for the same treatment trains applied to the same inventories, NRDI estimates are comparable. Thus, the NRDI estimates seem as reasonable as those from other sources.

There are three important limitations that make the NRDI total industry cost estimate lower than those from other sources. First, the NRDI excludes several industries from analysis. Second, the NRDI does not compute pretreatment costs for plants discharging directly to municipalities. Third, the control costs for thermal discharges are not computed. Table 3.12 compares NCWQ, EPA, and NRDI cost estimates.

COMPENSATION FOR UNDERESTIMATION OF COSTS

Residual discharge to municipal facilities varies considerably among industries: Some industries such as textiles and metal finishing discharge more than 50 percent of their residuals into them;

TABLE 3.12

Comparison of Capital Cost Estimates by NRDI, NCWQ, and EPA for 1973 Industry to Achieve
BPT and BAT
(millions of dollars)

Industry	BPT[a]			BAT[a]		
	NRDI	NCWQ	EPA	NRDI	NCWQ	EPA
In-Depth Study						
Pulp and paper	1,793	2,640	1,813	633	798	NA
Organic chemicals	1,341	3,346[b]	2,868	1,302	2,985[b]	NA
Petroleum refining	650	1,050	1,446	1,241	1,180	NA
Iron and steel	1,883	2,910	1,535	1,300	949	NA
Inorganic chemicals	544	520	419	139	247	NA
Plastics and synthetics	270	160	428	176	286	NA
Textile mills	600	537	117	322	300	NA
Steam/electric power	1,727	3,740	1,242	0	2,030	NA
Total (rounded)	8,800	14,900	9,900	5,000	8,800	NA
General Study						
Ore mining and dressing	74	610	74	0	0	NA
Coal mining	39	7,700	0	0	0	NA
Petroleum and gas extraction	138	234	120	0	1,070	NA
Mineral mining and processing	29	730	0	4	0	NA
Meat products and rendering	74	130	273	160	240	NA
Dairy products	52	79	402	93	51	NA
Grain mills	5	56	19	22	13	NA
Cane sugar processing	36	153	24	40	170	NA
Beet sugar processing	29	90	9	26	69	NA
Seafood	64	55	38	41	105	NA
Builders' papers	52	120	10	16	0	NA
Fertilizer	118	64	155	94	60	NA
Paving and roofing	4	6	5	2	4	NA
Rubber	240	220	105	86	48	NA
Leather tanning	89	119	67	16	73	NA
Glass	12	36	10	75	57	NA
Cement	68	34	28	0	9	NA
Pottery	29	3	18	20	4	NA
Concrete, gypsum	32	100	43	0	0	NA
Asbestos	3	4	5	3	9	NA
Ferroalloy	35	48	14	7	13	NA
Nonferrous metals	35	75	138	26	120	NA
Electroplating	260	14,140	194	65	14,100	NA
Fruits and vegetables	117	443	59	200	161	NA
Miscellaneous chemicals	1,323	944[b]	700[c]	751	655[b]	NA
Total	3,000	20,200	2,510	1,800	17,000	NA
Industries Not Studied in NRDI						
Fish hatcheries	NA	50	NA	NA	47	NA
Timber products	NA	14	70	NA	25	NA
Furniture and fixtures	NA	8	5	NA	0	NA
Paint and ink	NA	23	10	NA	0	NA
Soap and detergent	NA	11	19	NA	1	NA
Phosphates	NA	110	14	NA	26	NA
Structural clay	NA	5	NA	NA	0	NA
Fiberglass	NA	14	15	NA	0	NA
Transportation	NA	1,200	NA	NA	140	NA
Water supply	NA	1,200	1,000	NA	100	NA
Steam supply	NA	negative	NA	NA	negative	NA
Auto and other laundries	NA	25	NA	NA	21	NA
Foundries	NA	180	307	NA	0	NA
Nonferrous mill products	NA	260	7	NA	0	NA

TABLE 3.12 (Continued)

Industry	BPT[a]			BAT[a]		
	NRDI	NCWQ	EPA	NRDI	NCWQ	EPA
Miscellaneous food and beverages		5	NA	NA	5	NA
Machinery		3,900	1,404	NA	3,900	NA
Feedlots		2,205	58	NA		493	NA
Total		9,210	NA	NA	4,800	NA
Grand total		44,310			30,600	

[a]NRDI includes plants in 48 coterminous states; NCWQ and EPA include Alaska, Hawaii, and territories.

[b]NCWQ organic chemicals split into organic and miscellaneous chemicals.

[c]Estimated.

Note: NA means data not available.

Sources: NRDI Computer Outputs. 1975; Gianessi, L. P., and H. M. Peskin. 1975a. The Cost to Industries of Meeting the 1977 Provisions of the Water Control Amendments of 1972. Washington, D.C.: National Bureau of Economic Research; and National Commission on Water Quality. 1976. Staff Report to the National Commission on Water Quality. Washington, D.C.: Government Printing Office.

other industries such as iron and steel, and nonferrous metals discharge virtually none of their residuals into them.

Industrial residuals discharged into municipal facilities usually contain BOD and SS, which are compatible with municipal facilities' treatment capabilities. However, other industrial residual discharges are incompatible and regulations require pretreatment of these residuals before an industry can discharge into a municipal system. These incompatible residuals can have detrimental effects by damaging the treatment plants, reducing the performance of the system, or increasing the residuals discharge from the treatment plants.

Industrial dischargers bear two types of costs when they discharge into municipal systems. The first type of cost is a compensatory payment to the municipality for treating compatible industrial residuals. Industries are usually required to pay their share of capital costs and operation and maintenance costs based on flow and quantity of residuals. While these costs may be significant for some industries, they are not included in the NRDI industry costs of meeting the 1977 and 1983 objectives. The ST and BPWTT costs reflect the costs for both domestic and industrial waste flows in municipal plants. Thus, including the compensatory payment by industries to municipalities in meeting BPT and BAT objectives, costs would be double counting the costs to the nation for point-source residual reduction. The compensatory payments are best viewed as costs to industries. Assuming that industrial flow is approximately 20 percent of total municipal flow, the industrial share of incremental ST costs would be $5.7 billion and of incremental BPWTT costs would be $2.1 billion (see Chapter 4).

The second type of cost borne by industrial dischargers into municipal facilities are the technology costs for removing incompatible residuals such as acids, toxic substances, heavy metals, and nondegradable organics. Most of the industries studied in depth and many of those studied in general discharge incompatible residuals and must bear an additional technology cost if they remove them before they discharge into municipal plants. Since these costs are not included in either the ST and BPWTT or the BPT and BAT costs for direct dischargers, these costs are legitimate costs to the nation's industrial activities for application of BPT and BAT.

While these costs are legitimate costs to meet BPT and BAT requirements and should be included in analysis, they are very difficult to estimate for three reasons. First, EPA has not clearly defined the residuals considered incompatible with municipal plants. In regulations, EPA states that "the pretreatment standards themselves were not clearly understood by many segments of the general public" (EPA 1975). Second, many professionals in the field think that at least some so-called incompatible residuals, heavy metals, for example, are, in limited quantities, compatible and are effectively removed by municipal treatment facilities. If so, industries should not have to bear pretreatment costs for incompatible residuals and their only costs would be user charges. Third, the magnitude of current discharges of incompatible residuals cannot be estimated with much confidence. There is a scarcity of information on the number of establishments, by industrial category, discharging incompatible residuals and on the amount and type of residuals discharged per establishment. For example, the number of establishments in the metal finishing industry has been estimated to be 20,000 by EPA, and 70,000 by NCWQ. In light of these limitations, any estimation of pretreatment costs for industry is unlikely to be very accurate.

Since NCWQ estimated very large pretreatment costs for metal finishing, textiles, and fruits and vegetables, an attempt was made to estimate pretreatment costs for these three industries using the NRDI data base. These estimates are shown in Table 3.13.

The pretreatment costs for metal finishing were estimated by applying the BPT sequence of unit processes defined by EPA for electroplating industries to the amount of estimated flow, 85 percent of total industry flow in SICs 3471 and 3479, discharged into municipal facilities. The additional costs for the metal finishing industry for pretreatment of incompatible residuals are estimated to be about $1.5 billion. The pretreatment costs for textiles are estimated by applying BPT defined for each industry subcategory to the plants in each subcategory discharging to a municipality. It

TABLE 3.13

Pretreatment Costs for Selected Industries for BPT
(billions of dollars)

Industry	NRDI	NCWQ	EPA
Metal finishing	1.50	11.30	1.10
Textiles	0.20	00.40	0.04
Fruits and vegetables	0.07	00.20	0.00
Total	1.77	11.90	1.14

Sources: NRDI Computer Outputs. 1975; National Commission on Water Quality. 1976. Staff Report to the National Commission on Water Quality. Washington, D.C.: Government Printing Office; and Gianessi, L. P., and H. M. Peskin. 1975a. The Cost to Industries of Meeting the 1977 Provisions of the Water Pollution Control Act Amendments of 1972. Washington, D.C.: National Bureau of Economic Research.

TABLE 3.14

NRDI Adjusted Total Industry Cost Estimated to Apply BPT
(billions of 1975 dollars)

Category	BPT
NRDI industries	11.7
Pretreatment	1.8
Machinery and mechanical products	1.4
Water supply	1.1
Total	
NRDI	16.0
EPA	16.2
NCWQ	44.3

Sources: NRDI Computer Outputs. 1975; Gianessi, L. P., and H. M. Peskin. 1975a. The Cost to Industries of Meeting the 1977 Provisions of the Water Pollution Control Act Amendments of 1972. Washington, D.C.: National Bureau of Economic Research; National Commission on Water Quality. 1976. Staff Report to the National Commission on Water Quality. Washington, D.C.: Government Printing Office.

is estimated that 52 percent of the total industry flow is thus discharged. The costs for the textiles industry for pretreatment are
estimated to be about $0.2 billion. The pretreatment costs for the
fruits and vegetables industry are estimated to be about $0.07
billion.

In summary, industrial dischargers into municipalities bear
two types of costs, with one cost representing a user charge paid
to municipalities for treating compatible residuals. As Table 3.13
shows, the NRDI estimate for the latter is only 15 percent greater
than the NCWQ estimate, but is 155 percent of the EPA estimate.

Residuals Discharges and Costs for
Industries Excluded from NRDI

As mentioned earlier in this chapter, several industries were
not included in NRDI. Two of these industries—machinery and
mechanical products, and water supply—were estimated by NCWQ
to have high control costs. Machinery and mechanical products,
and water supply were not included in NRDI for two reasons. First,
neither was a major generator of BOD and SS. A brief analysis
using National Bureau of Economic Research (NBER) data (Gianessi
and Peskin 1975b) showed that, based upon 1968 discharge, BOD
and SS discharges from each of these two industries comprised
less than 1 percent of the respective national totals. Second, each
of these industries, particularly machinery and mechanical products, required a large research effort to analyze, and in light of
their relatively small BOD and SS discharges, the investment did
not seem warranted. Therefore, to approximate the control costs
for these industries, the EPA estimates shown in Table 3.14 are
used in NRDI.

In Gianessi and Peskin (1975a), it was estimated that for
existing sources in the machinery and mechanical products industry it will cost about $1.4 billion to meet BPT requirements. For
the water supply industry the estimate for existing sources was
about $1.1 billion to meet BPT. The costs for direct surface dis-
chargers to meet BAT requirements were not estimated in the
NBER report. Consequently, no approximation is made of the costs
for direct surface dischargers in these industries to meet BAT.

Table 3.14 shows the estimates made for pretreatment costs
and the costs of applying BPT in the major industries not included
in NRDI.

REFERENCES

Catalytic Inc. 1975. Capabilities and Costs of Technology for the
    Inorganic Chemicals Industry to Achieve the Effluent Limita-
    tions of P. L. 92-500. PB 244-800. Springfield, Va.:
    National Technical Information Service.

Environmental Protection Agency. 1973. Report to the Congress:
    The Economics of Clean Water. Washington, D.C.: Environ-
    mental Protection Agency.

_____. 1974a. Development Document for Major Organic Products.
    Washington, D.C.: Government Printing Office.

_____. 1974b. Development Document for Petroleum Refining.
    Washington, D.C.: Government Printing Office.

_____. 1974c. Technology Transfer: A Dry Caustic Method for
    Peeling Peaches. Washington, D.C.: Environmental Pro-
    tection Agency.

_____. 1975. Pretreatment Guidelines. Washington, D.C.: Environ-
    mental Protection Agency.

Gianessi, Leonard P., and Henry M. Peskin. 1975a. The Cost to
    Industries of Meeting the 1977 Provisions of the Water Pol-
    lution Control Amendments of 1972. Washington, D.C.:
    National Bureau of Economic Research.

_____. 1975b. Table of Water Effluent Discharges by SIC. Wash-
    ington, D.C.: National Bureau of Economic Research.

National Commission on Water Quality. 1976. Staff Report to the
    National Commission on Water Quality. Washington, D.C.:
    Government Printing Office.

National Research Council. 1975. National Residuals Discharge
    Inventory Computer Outputs. Washington, D.C.: National
    Research Council.

Pechan, Edward H. 1976. Program Documentation: The National
    Residuals Discharge Inventory. Washington, D.C.: National
    Research Council.

Preston, R.S., and Y.C. O'Brien. 1975. "The Wharton Long-term
    Annual and Industry Forecasting Model: Structural Character-
    istics and Policy Applications." Paper presented at the Con-
    ference on Mathematical Economic Models at Moscow, mimeo-
    graphed.

U.S. Department of Commerce, Bureau of the Census. 1971. 1967
    Census of Manufacturers: Water Use in Manufacturing.
    Special report series MC72(SR)-4. Washington, D.C.:
    Government Printing Office.

_____. 1972 Census of Manufacturers: Water Use in Manufacturing.
    Special report series MC72(SR)-4. Washington, D.C.: Gov-
    ernment Printing Office.

Water Resources Council. 1974a. 1972 OBERS Projections. Wash-
    ington, D. C. : Government Printing Office.
_____. 1974b. Series E OBERS Projections and Historical Data:
    Population, Personal Income, and Earnings—Aggregated
    Sub-areas. Washington, D. C. : Water Resources Council.
Wharton Econometric Forecasting Associates. 1975. Wharton
    Econometric Forecasting Estimate, Mark IV. Philadelphia.
    Wharton Econometric Forecasting Associates.

4

MUNICIPAL
ACTIVITIES

Municipal activities encompass residential, commercial, and industrial activities discharging residuals into publicly owned sewage treatment plants. The NRDI approach to municipal activities has been designed to simulate residual generation and discharges and costs of control technologies. The levels of technology application evaluated are:

1. The 1977 or ST technology goal, defined as those unit treatment processes meeting ST requirements as identified in the 1974 Joint State-EPA Survey of Needs (Needs Survey) (EPA 1975a).

2. The 1983 or BPWTT goal, defined as ST plus any additional unit processes more stringent than traditional ST processes requested in the Needs Survey; and

3. BPWTT+; a goal defined for each facility as ST plus either any additional unit processes more stringent than traditional ST processes requested in the Needs Survey or filtration where no additional process has been requested. This level is used to approximate a more stringent uniform treatment requirement after ST.

MUNICIPAL FACILITIES INVENTORY

The Needs Survey estimated the cost of construction needed by publicly owned treatment works to meet the 1983 goals of the 1972 Act. The cost reported (adjusted here to 1975 dollars) increased, from the prior (1973) survey, for all categories of needs from $78 billion to $445 billion, and for conventional treatment needs, treatment plants, and interceptors, from $47 billion to

$60 billion. The total increased significantly because of the expansion of the 1974 survey to include a category for treatment and control of stormwaters.

EPA recognized that the increase in the traditional treatment needs was a result not only of a better understanding of the requirements of the act on the part of the states, but also of the unevenness in the estimates. Some of the states "deviated significantly" from the EPA survey guidelines, according to the EPA final report (EPA 1975).

Given the questionable validity of the cost estimates in the Needs Survey as well as the difficulty in reaggregating its data, it was necessary to develop an NRDI approach to independently estimate the cost of meeting the 1977 and 1983 technological objectives. The NRDI also makes separate estimates of the cost for meeting ST and BPWTT objectives, presented as only one estimate in the Needs Survey. In addition, it provides data on residual generation and discharges, not reported in the Needs Survey.

The basic technique of the NRDI is to use the physical data regarding existing and projected facilities from the Needs Survey and to disregard cost data. The major steps in calculating these independent estimates of residuals and costs are illustrated in Figure 4.1 and explained below.

The first step (A in the figure) adjusts the survey data for omissions and inconsistencies. An adjustment is made for the sampling of existing plants serving less than 10,000 persons not in an SMSA. States were permitted by the survey to sample such plants rather than complete an individual form. A complete inventory for each state is generated by reproducing the sample records provided in the survey as many times as necessary to simulate full coverage.

The reported 1990 population-served estimate is checked with the Water Resources Council's OBERS series E 1990 state projections. If the 1990 state population-served estimate exceeds the OBERS state estimate, then the population to be served by future plants is reduced by 50 percent and, if necessary, the growth of future population to be served per plant is limited to the OBERS growth rate for the particular state.

Finally, a check on future needs of all plants is made. This involves either acceptance of specified unit treatment processes or an assignment of unit processes based on other survey data. A dollar request for ST is assumed to be activated sludge and a dollar request for tertiary treatment is assumed to be filtration.

The second step (B in the figure) specifies definitions of ST. BPWTT, and BPWTT+ goals. ST is defined as secondary treatment

FIGURE 4.1

Methodology for Municipal Facilities Analysis

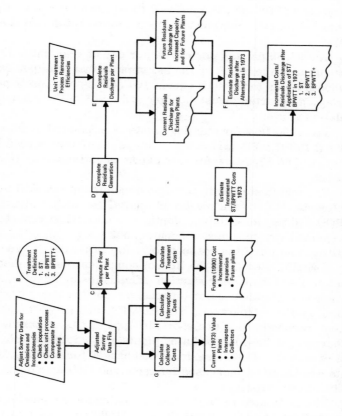

Source: Pechan, E. H. 1976. Program Documentation: The National Residuals Discharge Inventory. Washington, D. C.: National Research Council.

53

for all facilities. BPWTT is defined as any additional unit processes more stringent than the processes of the ST alternative. BPWTT+ is either any additional unit processes more stringent than ST requested in the Needs Survey or filtration if no additional process was requested.

The third step (C in the figure) computes wastewater flow for each plant. Although flow was reported in the survey for many plants, an estimate is computed for the sake of consistency. The 1973 flow in millions of gallons per day (MGD) is estimated by current flow (FC) = 97.85 x PC + F (i) where PC = current population served (in millions); and F (i) = reported industrial flow (in MGD).

The 97.85 gallons per capita per day of residential flow is a national average based on the assumption that total per capita (residential plus industrial) equals 120 gallons per capita per day, subtracting reported industrial flow and dividing the remainder by population served.

The 1990 flow for existing plants is estimated by future flow (FF) = (PF/PC x FC) where PF = future population served (in millions). The 1990 flow for new plants is estimated by (FF) = PF x 125.

The fourth step (D in the figure) computes residual generation or influents into sewage treatment plants. Residual influent data for BOD and SS, for existing plants, are used when reported. When not reported, the assumed concentrations were 200, 230, 40, and 10 milligrams per liter respectively, for BOD, SS, N, and P.

The fifth step (E in the figure) computes residual discharges by plant. Plant residual reduction efficiencies are derived from the sequence of unit treatment processes for the four alternatives studied; in-place treatment, ST, BPWTT, and BPWTT+. The unit process removals used are shown in Appendix C. Current residual discharges are based on existing plants and existing in-place treatment processes. Future residual discharges for existing and new plants are based on increased influents and the same or additional unit treatment processes.

The sixth step (F in the figure) estimates residual discharges after application of ST, BPWTT, and BPWTT+, based on 1973 population. The estimates are made by computing a percentage reduction for each residual by basin. These reductions are then applied to 1973 residual generation by basin; the discharges are defined to be residual discharges after ST, BPWTT, and BPWTT+ in 1973. This adjustment makes the initial municipal residual estimates comparable to the base-case industrial residual estimates, which assume no growth in production.

The seventh step (G in the figure) calculates costs of collector sewers for current and future conditions. Collector costs per capita for ten population-size groups are documented in Appendix C. The 1973 replacement value is based upon the population served by existing plants times the appropriate per capita cost. The 1990 collector costs are based upon the new population to be served times the per capita costs for the appropriate population-size group. The population requiring new collectors is estimated (when a nonzero cost for collectors was entered in the Needs Survey) by applying the percent increase in population to be served in that population-size group in the nation to the existing population for the facility.

The eighth step (H in the figure) calculates costs of interceptor sewers for current and future facilities. Interceptor costs by population-size group are accounted for by a factor multiplied by treatment plant capital costs. Factors are listed in Appendix C. The 1973 interceptor replacement values are based upon the capital value in place times the appropriate factor. Similarly, 1990 interceptor costs are based upon the incremental capital cost requirement.

The ninth step (I in the figure) calculates treatment costs for current and future conditions. Treatment costs are based upon flow and influent loads, and the appropriate unit treatment cost functions in Appendix C. Each facility requirement is defined as a treatment train of one or more unit treatment processes. The technique applies each of the specified unit processes sequentially and produces a total cost for the facility as a whole. Treatment costs for 1973 conditions are the replacement values of treatment facilities in place, and for 1990 conditions, are the incremental costs of additional treatment facilities at existing plants or the total costs of treatment facilities at new plants.

The tenth step (J in the figure) calculates incremental capital costs after the application of ST, BPWTT, and BPWTT+, based on 1973 flow. The computation involves multiplying 1990 costs by the ratio of 1973 flow to 1990 flow in each basin in order to arrive at 1973 cost. This adjustment makes the municipal cost estimates comparable to the initial industrial cost estimates, which assume no growth in production.

Municipal facilities for handling residential and industrial sewerage in 1973 totaled 24,209 plants (see Table 4.1). The largest size (MGD) category consists of 6 percent of the plants and accounts for 50 percent of the flow; the smallest consists of 90 percent of the plants and accounts for 11 percent of the flow. Minimal changes in this distribution will occur in 1990.

TABLE 4.1

Sewage Treatment Facilities

Size (MGD)	All Facilities in 1973		All Facilities in 1990	
	Number of Plants	Total MGD	Number of Plants	Total MGD
Less than 1*	21,822	2,313	21,186	2,737
1-5	1,794	3,831	2,181	4,817
5-20	447	4,187	646	6,046
More than 20	146	9,967	202	18,041
Total	24,209	20,298	24,209	31,641

*Includes facilities with no treatment flow.
Source: NRDI Computer Outputs. 1975.

TABLE 4.2

Municipal Residuals Generation, Discharge, and Costs

Projection Year/Treatment	Residuals (billion pounds per year)		Costs (billions of dollars)	
	BOD	SS	Treatment and Interceptor	Collector
1973/generation	13.2	14.1	NA	NA
1973/in-place	5.8	6.0	32.5	38.1
1973/ST	1.8	1.7	28.4	6.1
1973/BPWTT	0.9	0.8	10.3	6.1
1973/BPWTT+	0.5	0.4	10.0	0.0
1990/ST	2.8	2.7	35.1	7.8
1990/BPWTT	1.4	1.2	16.3	7.8
1900/BPWTT+	0.8	0.6	15.6	0.0

Source: NRDI Computer Outputs, 1975.

Municipal residuals generation is estimated to be 13.2 billion pounds per year of BOD and 14.1 billion pounds per year of SS (see Table 4.2). Investment in waste reduction facilities by 1973 had reduced the discharge to 5.8 billion pounds per year of BOD and 6.0 billion pounds per year of SS. This is approximately a 57 percent reduction of BOD and SS. The replacement value of the treatment facilities that resulted in this reduction is estimated to be $32.5 billion.

Also presented in Tables 4.1 and 4.2 are data on sewage treatment facilities, residual generation, and discharges projected for 1990, with total flow put at 31,600 MGD as shown.

GROWTH IN MUNICIPAL RESIDUALS GENERATION

Growth in residual generation by municipal facilities is influenced by several significant factors: population growth, population sewered, industrial discharge into municipal systems, and per capita loadings. In estimating the total 1990 flow, EPA instructed states to limit the population growth factor to the Census Bureau series E projection and to use a flow of 125 gallons per capita (EPA 1975a). An examination of the data submitted revealed a tendency for the projected total sewered population of a state to exceed the total series E 1990 projection. Consequently, adjustment techniques described earlier were applied.

The NRDI growth estimate for municipal facilities is derived from a two-stage process, described earlier. First, the 1990 population estimates per facility are adjusted, where necessary, to be compatible with state projections. Since the adjusted Needs Survey data are only for 1990 flow estimates, projections for other years are computed using interpolation as follows:

$$R(EST) = ((EST - 1973) \div 17) \times (R(1990) - R(1973)) + R(1973)$$

where: $R(EST)$ = residual estimation for year EST

$\qquad$ EST = target estimation year (for example 1985)

$\qquad$ $R(1990)$ = residuals in 1990

$\qquad$ $R(1973)$ = residuals in 1973

This technique assumes linearity in the growth from 1973 to 1990.
For example, the projection for 1985 is the 1973 estimate plus $\frac{12}{17}$ of the incremental change expected between 1973 and 1990.

LIMITATIONS OF MUNICIPAL STUDIES

NRDI provides data on municipal residuals and costs to meet
the various technology objectives. The NRDI data is preferable to
the EPA Needs Survey data and other sources because the residual
data are available for all facilities and the cost estimation technique well documented. However, this approach has its limitations
as discussed below.

The NRDI approach does not use the flow data in the Needs
Survey, which are available for many facilities and might provide
a more accurate estimate. However, there are many cases in
which the data are either not available or incorrect. For these
reasons, a consistent procedure for estimating flow for all facilities was used.

Municipal data for BOD and SS influent concentrations were
obtained from the Needs Survey, where reported, and were
estimated from national averages when unavailable. N and P data
were estimated in all cases. Because the municipal model used
specific influent data were possible, it theoretically represents the
most accurate type of data obtainable. In most cases, larger
facilities are able to submit accurate influent information, based on
the various influent testing programs required. The accuracy and
adequacy of data from smaller facilities was not tested, but has
been assumed to be less valid.

Variation in municipal loadings can be quite significant,
depending upon water supply source, infiltration/inflow problems,
use of garbage disposals, seasonal population, combined or separate
stormwater system, extent of industrial waste, and other factors.
Some municipal systems treat industrial wastes almost exclusively.
On the other hand, problems of excessive infiltration and inflow
can lead to a very dilute wastewater. Here again, research needs
to be done to relate these variables to the differing coefficients of
municipal residual generation.

Metcalf and Eddy, Inc. (1975) report a range from municipal
BOD of 100 to 300 milligrams per liter. Other sources cited in
their report show a BOD range of 75 to 390 milligrams per liter.
For SS, the reported values are 100 to 350 with other sources
showing 62 to 500 milligrams per liter. For phosphorus, the
reported range is from 6 to 20 milligrams per liter while for
nitrogen (as TKN), the range is 16 to 85 milligrams per liter.

NRDI COST ESTIMATES

NRDI may understate the costs of facilities construction in
specific instances, since application of statistical cost functions
does not reflect engineering difficulties, which could result in

TABLE 4.3

Comparison of State, EPA Adjusted, and NRDI Construction Cost Estimates
(billions of 1975 dollars)

Category	State Estimate[a]	EPA Adjusted Estimate[a]	NCWQ Estimate	NRDI Estimate[b]
1. Secondary treatment	16.9	16.9	10.8	15.6
2. More stringent treatment	26.0	20.8	24.8	20.8
4A. Collector sewers	32.5	22.1	13.0	15.6
4B. Interceptor sewers	26.0	23.4	13.5	15.0
Totals for 1, 2, and 4B	68.9	61.1	49.1	51.4
Totals for 1, 2, 4A, and 4B	101.4	83.2	62.6	67.0

[a]These estimates include Puerto Rico, Virgin Islands, and trust territories.
[b]This estimate excludes Puerto Rico, Virgin Islands, and trust territories.

Note: NRDI estimates were not made for Needs Survey categories 3A, 3B, 5, and 6 (infiltration, inflow, combined sewers, and stormwater treatment).

Source: NRDI Computer Outputs, 1975; Environmental Protection Agency. 1975a. Report to the Congress: Cost Estimates for Construction of Publicly Owned Treatment Facilities—1974 Needs Survey. Washington, D.C.: and National Commission on Water Quality. 1976.

actual costs very different from those estimated. However, on a national basis, such inaccuracies should average out.

The NRDI estimate differs from the NCWQ estimates and those reported in the 1974 Needs Survey (see Table 4.3). The NRDI estimate shows a reduced total for the four cost categories of $34 billion from the state estimate and $16 billion from the EPA adjusted estimate. However, the NRDI estimated value is $5 billion greater than the NCWQ value. In general, most of the difference in the estimates are in categories 4A and 4B, the sewer estimates. This is not surprising since it is more difficult to accurately estimate sewer costs than treatment plant costs.

REFERENCES

Environmental Protection Agency. 1975a. Report to the Congress: Cost Estimates for Construction of Publicly Owned Treatment Facilities—1974 Needs Survey. Washington, D. C.: Environmental Protection Agency.
_____. 1975b. Procedural Guidance: 1974 Joint State–EPA Survey of Needs for Municipal Wastewater Treatment Facilities. Washington, D. C.: Environmental Protection Agency.
Metcalf and Eddy, Inc. 1975. Report to the National Commission on Water Quality: Assessment of Technologies and Costs for Publicly Owned Treatment works under P. L. 92–500. PB 250 690. Springfield, Va.: National Technical Information Service.
National Commission on Water Quality. 1976. Staff Report to the National Commission on Water Quality. Washington, D. C.: Government Printing Office.
National Research Council. 1975. National Residuals Discharge Inventory Computer Outputs. Washington, D. C.: National Research Council.
Pechan, Edward H. 1976. Program Documentation: The National Residuals Discharge Inventory. Washington, D. C.: National Research Council.

5

Areal* activities include combined sewers, urban storm-water, agriculture, construction, mining, silviculture, and similar activities. The areal activities included in NRDI are limited to urban runoff (combined sewer overflows and urban stormwater) and nonirrigated agriculture. Residuals generation and discharge data for other agricultural operations and other activities are not available on a county-by-county basis and thus are not included in NRDI. However, it is known that in the case of residuals discharges from other major agricultural activities, the spatial impact is confined primarily to midwestern and western states.

Areal activities are distinguished from municipal and industrial activities for several reasons. First, areal activities are an intermittent rather than continuous source of residual discharges because their discharge is dependent upon rainfall events. Thus, there is considerable controversy about the impact of residuals from these activities on water quality, since increased stream flow associated with rainfall and snowmelt may dilute to some degree the impacts of residuals from these activities.

Second, areal activities are diffuse rather than concentrated sources. Some urban stormwater and most nonirrigated agriculture discharges directly enter surface waters rather than being

---

*Areal activities are defined as those representing wastewater collected across an area of land rather than as a production or consumption process.

collected and discharged at a limited number of points, again leading to a question about the impact of residuals from these activities
upon water quality.

Third, estimates of residuals and associated costs of technologies are taken either directly or in only a slightly modified
form from NCWQ contractor reports, whereas those for municipal
and industrial activities represent independent assessments.

Fourth, gross residuals data on areal sources are limited
and of questionable accuracy. The viability of the data are such
that SS estimates for nonirrigated agriculture could vary by as
much as plus or minus 200 percent for any river basin, and
estimates for urban runoff could vary by one or two orders of
magnitude. Thus, these estimates cannot be accepted with the
same confidence as those for point sources.

URBAN RUNOFF

The NRDI for urban runoff activities includes both combined
sewer overflows and urban stormwater events. The NRDI approach
estimates residual discharges from a specific storm event and
costs of controlling these discharges for urbanized areas within
SMSAs in the United States. Two sets of unit treatment processes
were selected: (level 1) screening, primary sedimentation, and
chlorination; and (level 2) these plus coagulation, secondary sedimentation, and filtration.

The method for analyzing gross and net residuals consists
of the following major steps (see Figure 5.1):

1. Identifying those counties in the United States in SMSAs
and with a population density equal to or greater than .6 people per
acre (there are 172 counties in the United States that fall into this
category);

2. Computing runoff and residual generation in combined-
sewer, separate-sewer, and nonsewered areas per county;

3. Computing residual reduction technology costs, and
interceptor and collector costs where appropriate, by county per
category for each treatment level;

4. Computing residual discharges by county per category
for each treatment level; and

5. Summing residual and cost estimates for appropriate geographic area—SMSAs, river basins, regions, or the nation.

FIGURE 5.1

Methodology for Urban Runoff Analysis

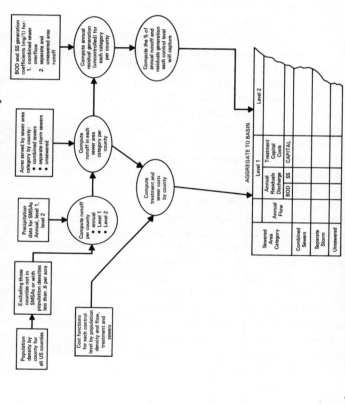

Source: Pechan, E. H. 1976. Program Documentation: The National Residuals Discharge Inventory. Washington, D.C.: National Research Council.

63

While the NRDI and NCWQ approaches are similar, one key difference is that NRDI calculates runoff for each county within SMSAs rather than each SMSA as a whole. This requires population density data for individual counties taken from the City-County Data Book (U.S. Department of Commerce 1974) and use of the 1974 EPA Needs Survey (EPA 1975) to determine the areas served by sewer systems. A comparison of the estimates of the area not served by combined sewers in each of the common 210 SMSAs with population densities of 0.6 or greater showed that estimates varied in 139 SMSAs by only $\pm$ 2 square miles and in 28 by $\pm$ 10 square miles.

Urban runoff activities generate a significant amount of BOD and SS residuals (see Table 5.1). The estimated annual gross BOD is 3.2 billion pounds and SS is 59.9 billion pounds. The areas not sewered account for 80 percent of the BOD; the areas served by combined sewers account for 6 percent. The 1973 net residuals from urban runoff are equal to gross residuals because there are considered to be no treatment facilities in place.

Since there are no effluent limitations for urban runoff, the NRDI removal efficiencies and costs are based on the previously stated technology assumptions. After the installation of level 1 technologies, net BOD from urban runoff would decrease to 2.2 billion pounds annually or 69 percent of uncontrolled conditions, and net SS would decrease to 27.1 billion pounds annually or 45 percent of uncontrolled conditions. The capital costs of level 1 technology, exclusive of sewer costs, are estimated to be $62 billion. After the installation of level 2 unit processes, net BOD would decrease to 0.4 billion pounds annually or 13 percent of the uncontrolled conditions, and net SS would decrease to 3.5 billion pounds annually or 6 percent of uncontrolled conditions. The capital costs of level 2 technology, exclusive of sewer costs, are estimated to be $212 billion, or 414 percent higher than the level 1 costs. Installation of level 2 technology for urban runoff would result in a decrease of 82 percent in net BOD and 87 percent in net SS of level 1 quantities.

Limitations of Methodology

Although the storm events method used in NRDI makes possible a straightforward estimation of treatment facility costs, the results of such a formulation suffer from two limitations. One is the difficulty of estimating the proportion of total annual runoff captured by a treatment system designed for a specific storm event. The second is the difficulty of estimating gross residuals, given

TABLE 5.1

Urban Runoff Residuals Generation, Discharge, Treatment
Capital Costs, and Sewer Capital Costs

Type of Area	Control Level	Annual Residuals Generation			Treatment Capital Costs (billions of dollars)	Annual Residuals Discharge after Modification	
		Runoff (trillions of gallons)	BOD (billions of pounds)	SS (billions of pounds)		BOD (billions of pounds)	SS (billions of pounds)
Combined-sewer	1	0.6	0.4	2.8	8.9	0.29	1.30
Combined-sewer	2	0.6	0.4	2.8	26.0	0.06	0.11
Separate-sewer	1	0.9	0.2	4.0	4.3	0.14	1.80
Separate-sewer	2	0.9	0.2	4.0	14.7	0.02	0.24
Not sewered	1	12.1	2.6	53.0	38.1	1.80	24.00
Not sewered	2	12.1	2.6	53.0	172.0	0.31	3.10
Total	1	13.6	3.2	59.9	51.2	2.20	27.10
Total	2	13.6	3.2	59.9	212.0	0.39	3.50

Sewer Capital Costs (billions of dollars)

	Collector	Interceptor	Total
Combined-sewer	NA	1.5	1.5
Separate-sewer	NA	–	0.0
Not sewered	73	–	73.0
Total	73	1.5	74.5

Note: Dash means costs not computed.
Source: NRDI Computer Outputs. 1975.

different surface conditions and variable intervals between rainfall events.

A 1975 study (J. Heaney, W. Huber, and M. Murphy) addressed the first limitation. The study related storage and treatment rates for six cities falling in five precipitation regions of
the United States. This documented relationship was used to
estimate what proportion of annual runoff would be captured for
each storm event. However, the results of this exercise proved
inconclusive. For four out of the six cities, the capture of annual
runoff was only 80 percent, and for the remaining two, only 90
percent. Given the technology applied in the NRDI analysis, the
percent of annual runoff captured, for storm events criteria of
one year and one hour, exceeded the uppermost isoquant, that
is, was greater than either 80 percent or 90 percent, respectively,
indicating that the proportion of annual runoff captured in each of
the two storm events varies, from at least 80 percent (probably
90 percent) to 100 percent. In light of this, a 92 percent capture
was assumed for storm events criteria of two years and one hour,
with use of technology level 1, and a 98 percent capture for events
criteria of one year and 24 hours, with use of technology level 2.

Considering the second limitation, insufficient empirical
research has been performed to enable an accurate estimation of
gross residuals. The method used here (assigning a nationwide
mean concentration) is, clearly, a major simplification. Most
researchers agree that gross BOD and SS discharges do not
correlate linearly with runoff, particularly in the case of combined
sewer overflow. The concentration ranges are quite large.

One figure that may be analyzed for sensitivity is the concentration of BOD in storm water runoff. For separate sewer
systems, the NCWQ contractor reported a mean BOD of 26.6
milligrams per liter. The range for city averages was from 6 to
137 milligrams per liter, and the range for extremes was very
large. For the 11 cities surveyed, the low end of the range was
from 1.7 to 11.8 times the mean. The largest range for a specific
city was from 4 to 433 milligrams per liter, a difference of over
100 times.

Changes in the concentration of BOD in urban runoff would
obviously have impact on the relationship of runoff BOD to other
BOD. Among the cities surveyed, the low end of the range for
BOD is about 20 percent of the mean, and the high end of the range
is about 3.5 times the mean. If these factors were applied to the
current BOD generated and discharged, they would demonstrate
reasonable limits of the potential variability of concentration of
BOD in urban runoff. In Table 5.2 the mean figures for 1973

TABLE 5.2

Effects of Concentration Variability on BOD in Urban Runoff
(billions of pounds per year)

Range and Type of BOD	1973	BPT	BAT
Mean			
Urban runoff	3.3	3.3	3.3
Point-source	10.1	2.7	1.3
Runoff as percent of point-source	33.0	122.0	254.0
High end			
Urban runoff	11.6	11.6	11.6
Point-source	10.1	2.7	1.3
Runoff as percent of point-source	115.0	430.0	892.0
Low end			
Urban runoff	0.7	0.7	0.7
Point-source	10.1	2.7	1.3
Runoff as percent of point-source	7.0	26.0	54.0

Source: NRDI Computer Outputs. 1975.

discharge show that urban runoff BOD is 33 percent of the combined point-source BOD for industrial and municipal activities. Through BAT/BPWTT, the dramatic decline in point-source BOD means that runoff BOD would represent 254 percent of point-source BOD. The high end of the range (times 3.5) would produce 11.6 billion pounds of BOD per year, increasing from 115 percent of point-source BOD in 1973 to 892 percent at BAT/BPWTT. However, at the low end of the range runoff BOD would be approximately equal to point-source BOD, and the 0.7 billion pounds per year would amount to 54 percent of point-source BOD at BAT/BPWTT.

The general lack of information on urban runoff makes it more likely that present data underestimate BOD concentration rather than overestimate, so it is more likely that runoff BOD would be between the mean and the high end of the range rather than toward the lower end. This consideration of BOD may overlook an even more serious problem, the high amounts of heavy metals in urban runoff, estimated at perhaps 50 million pounds per year. Those heavy metals that make it to a sewage treatment system may be largely removed, but the overflow problem and dispersion of runoff means that much of this heavy metal would not be removed.

FIGURE 5.2

Methodology for Nonirrigated Agriculture Analysis

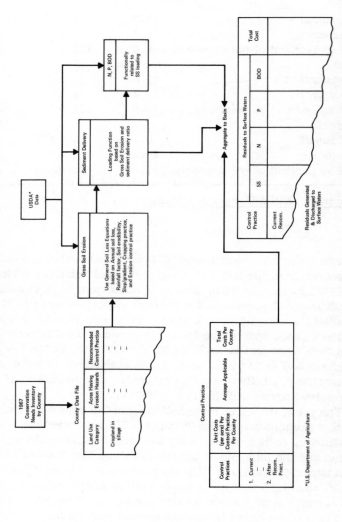

Source: Midwest Research Institute. 1974. User's Handbook for Assessment of Water Pollution from Nonpoint Sources. Washington, D. C. : Environmental Protection Agency.

NONIRRIGATED AGRICULTURE

For the nonirrigated agriculture category the NRDI is an adjusted version of the inventory prepared for NCWQ by Midwest Research Institute (MRI) (1976). An analysis of sediment loadings, using MRI and Iowa State University (1975) sediment delivery ratios, showed that, in areas where measured instream suspended sediment data existed, sediment loadings obtained using Iowa State ratios appeared to be more accurate than those obtained using MRI ratios. Both Iowa State and MRI use the U.S. Department of Agriculture's 1967 Conservation Needs Inventory and the universal soil loss equation to estimate gross soil erosion. Consequently, the residual loadings estimated by MRI are adjusted based upon the Iowa State delivery ratios. This method is illustrated in Figure 5.2.

To compute sediment discharge, MRI applied the universal soil loss equation to estimate gross soil erosion. They then approximated the amount of eroded soil entering surface waters by applying a sediment delivery ratio. NRDI adjusts the sediment loading estimates by multiplying them by the ratio of the Iowa State sediment delivery ratio divided by the MRI ratio for each basin. Since all other residual discharges are assumed to be multiplicative functions of sediment loading, all residual loadings are adjusted.

The functions used are based upon residual (nutrient) content in soil and sediment yields. The assumption is that the majority of these residuals are removed with the eroded sediment. A more complex discussion of these functions was prepared by MRI (1974).

Based upon the 1967 Conservation Needs Inventory, MRI computed residual discharges from nonirrigated agriculture by county for acres of different land classifications. This estimate of residual discharge (adjusted) is defined as "current control conditions." Reduction in residual discharges is estimated for only one control level. This level assumes the implementation of the control practices recommended in the Conservation Needs Inventory for acreage with erosion hazards.

The costs (MRI cost estimates are not adjusted) for implementing the control practices recommended in the Conservation Needs Inventory are calculated using unit costs, taken from the national Agriculture Conservation Program, of each conservation practice for each county. All county costs are then summed at the river basin level. Reporting on sediment from nonirrigated cropland, and the cost of control practices, MRI noted:

Cropland in the United States in 1969 was estimated to be 482.7 million acres. The acreage of non-irrigated cropland in tillage rotation is 378.4 million

TABLE 5.3

Gross and Net Residuals from Nonirrigated Agriculture,
and Control Costs

Source for Sediment Delivery Ratios	Control Level	Residuals (billions of pounds per year)				Capital Costs (billions of dollars)
		BOD	SS	N	P	
NCWQ	None	17.1	3,599.5	8.6	3.0	0.0
NCWQ	Controlled*	8.8	1,843.4	4.4	1.6	2.1
NRDI	None	6.5	1,373.8	3.3	1.3	0.0
NRDI	Controlled*	3.3	700.4	1.7	0.7	2.1

*Per Conservation Needs Inventory as discussed in text.
Sources: Midwest Research Institute. 1976. Water Pollution
Abatement Technology: Capabilities and Costs, Control of Water
Pollution from Selected Nonpoint Sources; and NRDI Computer
Outputs. 1975.

acres. About 40 percent of this land (158 million
acres) was listed as needing conservation treatment.
The total amount of sediment produced from nonirrigated cropland was estimated to be about 1,800
million tons per year. This is equivalent to about
4.8 tons/acre/year on an average basis. If needed
conservation practices were implemented, the sediment delivered would be reduced to 922 million tons
per year, or about 2.44 tons/acre/year.
The total investment cost of conservation treatment
was estimated to be $2.1 billion. The annualized
cost would amount to $500 million per year, or
$3.10 acre/year. The cost of control is accordingly
about $.75/ton (MRI 1976).

Information on gross and net residuals and costs is displayed
in Table 5.3.

Limitations of Methodology

In measuring concentrations of suspended solids, the data
base used, the 1967 Conservation Needs Inventory, is recognized

as the best national data base of its type presently available. The
major uncertainty is the portion of gross soil erosion delivered to
surface waters. Although Iowa State sediment delivery ratios produce, in the test cases used, better estimates of sediment loading
than do MRI ratios, the large discrepancies between them, generally about 100 percent, indicate that loading estimates may
contain considerable error. If a 100 percent error may be expected
for sediment loadings, then at least a 100 percent error would also
be expected for other residual loadings since they are estimated
based on sediment loadings.

The county soil concentrations of phosphorus are based upon
a 1946 U.S. Department of Agriculture map of phosphoric acid in
the top one foot of soil in the United States. In addition to map
reading inaccuracies, the 1946 phosphorus soil concentrations may
be quite different than present concentrations, given the large increase in fertilizer use.

The nitrogen content in soil is estimated using an empirical
expression based upon temperature and humidity (Jenny 1930). The
basic data (temperature, precipitation, vapor pressure, humidity)
necessary to use this expression are fairly reliable and county
specific. However, the expression seems to be estimating nitrogen
soil content for an undefined "natural" soil condition. As in the case
of phosphorus loading, a seemingly significant variable, fertilizer
application rate, is omitted from the analysis.

BOD soil concentration estimates are based upon nitrogen
concentrations. The assumption is that BOD concentrations are
uniformly twice those of nitrogen. Thus, BOD loading estimates
are, at best, only as good as those for nitrogen.

Sensitivity of data is shown by comparing results using NCWQ
sediment delivery ratios against those using NRDI values. This is
shown in Table 5.3.

REFERENCES

Environmental Protection Agency. 1975. Procedural Guidance:
    1974 Joint State-EPA Survey of Needs for Municipal Waste-
    water Treatment Facilities. Washington, D.C.: Environ-
    mental Protection Agency.
Heaney, J., W. Huber, and M. Murphy. 1975. "Nationwide
    Assessment of the Cost of Controlling Pollution from Com-
    bined Sewer Overflows and Stormwater Runoff From Sewered
    and Non-Sewered Urban Areas." Gainesville: University of
    Florida, Department of Environmental Engineering Sciences,
    mimeographed.

Iowa State University. 1975. "A National Model for Sediment
    Water Quality and Agricultural Production." Center for
    Agricultural and Rural Development report. Ames, Iowa,
    mimeographed.
Midwest Research Institute. 1974. User's Handbook for Assess-
    ment of Water Pollution from Nonpoint Sources. Washington,
    D. C. : Environmental Protection Agency.
_____. 1976. Water Pollution Abatement Technology: Capabilities
    and Costs for Control of Water Pollution from Selected Non-
    point Sources. Springfield, Va. : National Technical Infor-
    mation Service.
National Research Council. 1975. National Residuals Discharge
    Inventory Computer Outputs. Washington, D. C. : National
    Research Council.
Pechan, Edward H. 1976. Program Documentation: The National
    Residuals Discharge Inventory. Washington, D. C. : National
    Research Council.
U.S. Department of Commerce, Bureau of the Census. 1974. City
    County Data Book. Washington, D. C. : Government Printing
    Office.

6

**WATER
QUALITY**

Preceding chapters have explained the development of a set
of activity inventories that together produce a relatively complete
picture of gross and net residuals in each of the basins of the
contiguous United States. However, knowledge of residuals discharges alone does not permit an estimate of water quality, improvement of which is the ultimate objective of any control program.
Even if the residuals quantities are relatively small, there may be
severe water quality problems resulting from discharges into stream
segments with minimal flow. Unfortunately, water quality inter-
pretations relating to such local conditions are currently impossible
to analyze on a national basis.

The purpose of the analysis presented here is to rank the 99
river basins according to some relative index of average water
quality. Such a ranking may then be used to identify those basins of
the nation that are relatively well off or those that have problems
under current conditions, as well as changes due to different water
quality management policies. Because average conditions are determined for each basin and only two parameters are used, the analysis
may not reflect a real situation at any given location or time and may
not be used to pinpoint specific water quality problems in subbasins
or stream segments. Nevertheless, the relative ranking provides
one basis for allocating water quality management efforts.

WATER QUALITY INDEX

A number of computational aids and models are available
to predict or estimate water quality as affected by discharges of

residuals. Data requirements for these systems consist of the quantities, locations, and types of residuals being discharged into the waterway, and physical information describing the hydrology for the area under study.

The residuals data needed for the water quality model are derived from the activity inventories, with the additional requirement of more specific siting information. Hydrologic data, which vary depending upon the type of water quality estimation to be made, often include such information as stream flows, channel geometry, temperature conditions, and residuals reaction coefficients.

Water quality parameters can be further subdivided into those that measure concentrations of residuals (such as metals or toxic substances), and those that indirectly indicate the effects of residuals (dissolved oxygen, pH, fish biomass). Some residuals are conservative, meaning that they do not undergo chemical change after they enter a watercourse but are carried along with the flow. These residuals are easier to study than nonconservative residuals, which are acted upon and changed in the stream system.

An example of how a real basin is modeled with a sophisticated water quality model and how an actual basin is ranked in the NRDI illustrates the limited scope of the NRDI analysis. In Figure 6.1, diagram A, the real system, consists of a river system with a main stem and tributary systems. At various locations along the network, point and areal discharges enter the waterway, affecting the water quality where they enter and for some distance downstream.

The simplified model shown in diagram B can be used to measure the relative concentration of residuals in the basin at the basin exit point only. This is computed as follows:

$$C(X) = L \div Q \tag{6.1}$$

where: $C(X)$ = the concentration measure over the parameter
            set X
      L = the residual loading set for the entire basin
      Q = the total flow (stream or wastewater flow, whichever is greater) measured at the basin discharge point

The model in diagram B will accurately reflect concentrations of conservative residuals at the discharge point of the basin. However, it cannot estimate as accurately the concentrations of nonconservative residuals.

FIGURE 6.1

Real and NRDI Model Basins

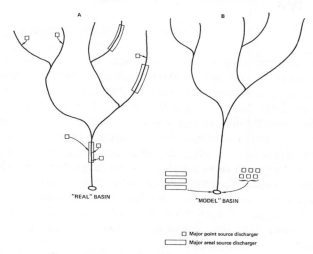

Source: Compiled by the authors.

Other attempts have been made to estimate water quality
in large areas. One pioneering effort by Wollman and Bonem
(1971) consisted of a model to measure trade-offs between installation of wastewater treatment facilities and construction of
reservoir facilities to augment low flows (and thus assimilation
capacity of wastes).

The analysis in the Wollman-Bonem model was based on
18 "major" basins. The model used average hydrological characteristics (such as flow, stream velocity, and reaction coefficients) and two categories of residual loadings; point sources
within SMSAs, and areal sources and point sources outside SMSAs.
Water quality measurements for dissolved oxygen, nutrients, and
thermal loadings were then computed by assuming that SMSA point-
source loads are discharged into one place in the stream and all
other sources are discharged evenly along all the stream miles in
the basin.

The procedures used in the Wollman-Bonem model are quite
complex and based on many different assumptions. While many of
the data inaccuracies can be averaged out when studied over as
large an area as a major basin, the problems would become more
acute if smaller basins were to be considered.

The results of the analysis indicated that a dilution of residuals was a feasible and sometimes cost-effective method of meeting
water quality standards.

NRDI Water Quality Ranking Procedure

The water quality ranking procedure developed as part of
the NRDI effort to account for the differential effects of point and
areal sources and stream flow on water quality. It involves making dilution estimates of both BOD and SS, during low and average
stream flow conditions. Under low flow conditions, discharges of
residuals from areal sources—urban storm runoff, and nonirrigated agriculture—are negligible, and only point-source discharges
are used in the low flow estimate. For average flow conditions,
discharges of residuals from both point and areal sources are
included. Thus, under low flow conditions, L in equation 6.1
reflects municipal and industrial discharge only. Under average
flow conditions, L is the sum of all residuals discharge in the
river basin.

The flow Q in equation 6.1 represents only flow discharge
from a single point in most basins and from multiple points in
basins on the coastal areas or in the Great Lakes. For these
areas, however, the flow data do not represent the actual volume
of water in the oceans or Great Lakes and thus do not reflect the
actual dilution capacity for those basins with coastal areas. In
order to compensate for this limitation, residual discharges from
coastal counties and counties adjacent to the Great Lakes are
excluded from the basin totals when computing water quality estimates. The coastal counties, defined by the Office of Coastal Zone
Management, U.S. Department of Commerce, are those counties
where surface waters are subject to tidal action (see Figure 6.2).

The actual flow measures may not always be the flow measure used in the dilution calculation. In less than 10 basins, the
stream flow is less than the flow from point sources, which results
in the concentration in the stream being greater than the concentration in the waste flow. These basins are in the Great Basin and
California-South Pacific regions. In basins where the point-source
flow is greater than the stream flow, the point-source flow is taken
as the flow for the water quality estimate.

The results of the dilution analysis for BOD and SS are presented for 1973 conditions for the 99 basins in Tables 6.1 and 6.2,
respectively. For the BOD dilution rankings, the concentrations
of BOD in milligrams per liter are listed for low and average

FIGURE 6.2

Coastal Counties Excluded from Water Quality Index

Source: U.S. Department of Commerce. 1975. List of Coastal Counties. Office of Coastal
Zone Management, mimeographed.

TABLE 6.1

BOD Dilution Rankings by Basin for 1973 Conditions
(lowest to highest BOD concentration, in milligrams per liter)

	Low Flow Conditions				Low Flow Conditions		
Rank	Basin Number*	Region Number*	Concen-tration	Rank	Basin Number*	Region Number*	Concen-tration
1	1707	17	0.0	51	102	1	5.1
2	1003	10	0.0	52	501	5	5.2
3	1001	10	0.0	53	1105	11	5.5
4	1403	14	0.1	54	306	3	5.6
5	802	8	0.2	55	303	3	5.8
6	801	8	0.2	56	404	4	5.9
7	803	8	0.3	57	1706	17	6.2
8	1702	17	0.3	58	502	5	6.3
9	1101	11	0.3	59	205	2	6.9
10	1704	17	0.4	60	1103	11	6.9
11	1401	14	0.4	61	302	3	7.0
12	602	6	0.4	62	506	5	7.1
13	1004	10	0.4	63	1201	12	7.2
14	505	5	0.5	64	105	1	7.7
15	1002	10	0.6	65	2105	12	7.7
16	106	1	0.7	66	1305	13	8.4
17	704	7	0.8	67	402	4	8.5
18	1705	17	0.8	68	201	2	9.5
19	1005	10	0.9	69	901	9	10.0
20	101	1	0.9	70	701	7	10.4
21	1006	10	1.1	71	1106	11	10.4
22	1402	14	1.1	72	1604	16	11.4
23	705	7	1.2	73	1503	15	12.0
24	1011	10	1.2	74	1304	13	12.9
25	309	3	1.5	75	104	1	15.3
26	507	5	1.5	76	1803	18	15.6
27	1701	17	1.7	77	204	2	16.6
28	405	4	1.7	78	206	2	17.2
29	1303	13	1.8	79	503	5	17.7
30	1603	16	2.1	80	203	2	17.7
31	1301	13	2.1	81	1203	12	18.3
32	308	3	2.2	82	1204	12	25.3
33	1008	10	2.6	83	1204	12	25.3
34	1801	18	2.7	84	1501	15	29.5
35	301	3	2.7	85	202	2	> 30.0
36	702	7	2.8	86	305	3	> 30.0
37	1104	11	2.9	87	1602	16	> 30.0
38	1703	17	3.1	88	1202	12	> 30.0
39	703	7	3.2	89	1302	13	> 30.0
40	401	4	3.5	90	1103	11	> 30.0
41	601	6	3.7	91	103	1	> 30.0
42	1802	18	3.8	92	1007	10	> 30.0
43	1009	10	4.1	93	406	4	> 30.0
44	408	4	4.2	94	407	4	> 30.0
45	1107	11	4.2	95	403	4	> 30.0
46	307	3	4.4	96	1807	18	> 30.0
47	1502	15	4.6	97	1805	18	> 30.0
48	1010	10	4.6	98	1804	18	> 30.0
49	504	5	4.7	99	1806	18	> 30.0
50	304	3	4.7				

	Average Flow Conditions				Average Flow Conditions		
Rank	Basin Number	Region Number	Concen- tration	Rank	Basin Number	Region Number	Concen- tration
1	803	8	0.3	51	1001	10	10.6
2	1403	14	0.3	52	302	3	11.2
3	1801	18	0.4	53	402	4	12.7
4	1705	17	0.4	54	704	7	12.8
5	101	1	0.6	55	1002	10	13.0
6	1301	13	0.8	56	1302	13	13.5
7	1303	13	0.8	57	305	3	13.9
8	1402	14	0.9	58	404	4	14.1
9	1703	17	1.1	59	1105	11	14.8
10	802	8	1.2	60	1005	10	15.8
11	1401	14	1.3	61	702	7	16.4
12	1701	17	1.3	62	204	2	16.7
13	1003	10	1.5	63	601	6	17.4
14	1802	18	1.9	64	1011	10	19.2
15	102	1	1.9	65	1205	12	19.3
16	106	1	2.0	66	1501	15	19.5
17	401	4	2.0	67	202	2	22.3
18	1704	17	2.0	68	203	2	22.7
19	801	8	2.2	69	206	2	23.4
20	1502	15	2.4	70	503	5	24.3
21	505	5	2.4	71	1202	12	27.0
22	1703	17	2.6	72	1006	10	27.1
23	1803	18	2.8	73	1805	18	27.3
24	1107	11	2.9	74	407	4	> 30.0
25	308	3	3.0	75	1203	12	> 30.0
26	705	7	3.1	76	506	5	> 30.0
27	1201	12	3.2	77	406	4	> 30.0
28	1101	11	3.4	78	1707	17	> 30.0
29	1706	17	3.4	79	103	1	> 30.0
30	1004	10	4.3	80	1204	12	> 30.0
31	501	1	4.3	81	1804	18	> 30.0
32	1304	13	5.0	82	1009	10	> 30.0
33	105	1	5.0	83	703	7	> 30.0
34	502	5	5.1	84	901	9	> 30.0
35	309	3	5.5	85	1103	11	> 30.0
36	304	3	6.0	86	1106	11	> 30.0
37	602	6	6.2	87	701	7	> 30.0
38	205	2	6.3	88	1007	10	> 30.0
39	405	4	6.3	89	1008	10	> 30.0
40	303	3	6.5	90	1102	11	> 30.0
41	307	3	6.5	91	1010	10	> 30.0
42	408	4	6.7	92	1807	18	> 30.0
43	306	3	7.3	93	403	4	> 30.0
44	504	5	7.4	94	1603	16	> 30.0
45	201	2	7.8	95	1806	18	> 30.0
46	1104	11	8.3	96	1503	15	> 30.0
47	301	3	8.4	97	1604	16	> 30.0
48	507	5	8.4	98	1602	16	> 30.0
49	1305	13	9.9	99	1601	16	> 30.0
50	104	1	10.4				

*As displayed in Figure 2.2 and defined in Appendix B.
Source: NRDI Computer Outputs. 1975.

TABLE 6.2

SS Dilution Rankings by Basin for 1973 Conditions
(lowest to highest SS concentration, in milligrams per liter)

	Low Flow Conditions				Low Flow Conditions		
Rank	Basin Number*	Region Number*	Concen- tration	Rank	Basin Number*	Region Number*	Concen- tration
1	1707	17	0.0	51	901	9	32.2
2	1001	10	0.0	52	1203	12	33.2
3	1801	18	0.2	53	502	5	34.8
4	1704	17	0.4	54	1201	12	40.2
5	803	8	0.5	55	408	4	41.2
6	1705	17	0.7	56	1010	10	41.8
7	1003	10	0.7	57	506	5	43.2
8	801	8	0.9	58	1303	13	43.3
9	802	8	1.1	59	204	2	44.8
10	704	7	1.6	60	203	2	51.8
11	304	3	1.6	61	1401	14	52.0
12	505	5	2.0	62	103	1	55.1
13	1802	18	2.4	63	206	2	61.8
14	1011	10	2.5	64	402	4	62.8
15	1101	11	2.9	65	1501	15	63.0
16	1006	10	3.5	66	1004	10	65.8
17	1706	17	3.5	67	305	3	71.0
18	602	6	3.7	68	1702	17	71.9
19	301	3	4.0	69	1301	13	75.2
20	102	1	4.2	70	1204	12	81.9
21	1009	10	4.8	71	1106	11	>100.0
22	1104	11	4.8	72	1202	12	>100.0
23	507	5	5.0	73	1205	12	>100.0
24	307	3	6.0	74	1102	11	>100.0
25	309	3	6.4	75	601	6	>100.0
26	205	2	6.5	76	503	5	>100.0
27	705	7	7.0	77	403	4	>100.0
28	1305	13	7.3	78	406	4	>100.0
29	702	7	7.4	79	701	7	>100.0
30	303	3	7.8	80	202	2	>100.0
31	201	2	9.7	81	1701	17	>100.0
32	1403	14	10.8	82	1402	14	>100.0
33	101	1	10.9	83	1103	11	>100.0
34	308	3	10.9	84	1805	18	>100.0
35	1008	10	11.2	85	407	4	>100.0
36	1803	18	11.7	86	1502	15	>100.0
37	302	3	12.6	87	401	4	>100.0
38	1105	11	14.4	88	1304	13	>100.0
39	703	7	14.7	89	1007	10	>100.0
40	1703	17	16.9	90	1804	18	>100.0
41	1107	11	17.1	91	1005	10	>100.0
42	306	3	17.4	92	1602	16	>100.0
43	504	5	17.7	93	1806	18	>100.0
44	104	1	18.6	94	1807	18	>100.0
45	405	4	18.8	95	1601	16	>100.0
46	1002	10	19.3	96	1302	13	>100.0
47	404	4	18.8	97	1604	16	>100.0
48	106	1	22.0	98	1503	15	>100.0
49	105	1	22.8	99	1603	16	>100.0
50	501	5	24.8				

	Average Flow Conditions				Average Flow Conditions		
Rank	Basin Number	Region Number	Concen- tration	Rank	Basin Number	Region Number	Concen- tration
1	1706	17	1.4	51	601	6	>100.0
2	102	1	3.1	52	406	4	>100.0
3	205	2	7.8	53	1001	10	>100.0
4	101	1	9.7	54	1107	11	>100.0
5	305	3	12.8	55	1104	11	>100.0
6	1305	13	13.3	56	1205	12	>100.0
7	304	3	14.1	57	506	5	>100.0
8	1705	17	14.2	58	1003	10	>100.0
9	1803	18	14.3	59	704	7	>100.0
10	1301	13	28.4	60	503	5	>100.0
11	303	3	30.5	61	602	6	>100.0
12	802	8	30.9	62	1105	11	>100.0
13	201	2	32.5	63	407	4	>100.0
14	104	1	34.3	64	703	7	>100.0
15	803	8	34.8	65	1804	18	>100.0
16	301	3	36.2	66	1002	10	>100.0
17	801	8	37.5	67	1701	17	>100.0
18	1201	12	41.0	68	1101	11	>100.0
19	306	3	41.6	69	1204	12	>100.0
20	308	3	47.5	70	401	4	>100.0
21	507	5	51.4	71	403	4	>100.0
22	408	4	54.2	72	1011	10	>100.0
23	106	1	56.6	73	1801	18	>100.0
24	405	4	57.9	74	1009	10	>100.0
25	302	3	58.3	75	1106	11	>100.0
26	307	3	60.2	76	1203	12	>100.0
27	1006	10	61.2	77	1103	11	>100.0
28	105	1	70.0	78	1402	14	>100.0
29	1802	18	70.1	79	1502	15	>100.0
30	501	5	72.7	80	1008	10	>100.0
31	1703	17	72.9	81	1403	14	>100.0
32	204	2	82.7	82	1403	14	>100.0
33	309	3	84.6	83	1303	13	>100.0
34	402	4	85.8	84	1010	10	>100.0
35	1702	17	90.0	85	1304	13	>100.0
36	206	2	90.9	86	1401	14	>100.0
37	702	7	92.6	87	1007	10	>100.0
38	203	2	95.7	88	1102	11	>100.0
39	1805	18	100.0	89	1806	18	>100.0
40	502	5	100.0	90	1005	10	>100.0
41	701	7	100.0	91	1302	13	>100.0
42	404	4	100.0	92	1501	15	>100.0
43	1704	17	100.0	93	1807	18	>100.0
44	202	2	100.0	94	1707	17	>100.0
45	505	5	100.0	95	1602	16	>100.0
46	103	1	100.0	96	1503	15	>100.0
47	504	5	100.0	97	1604	16	>100.0
48	705	7	100.0	98	1601	16	>100.0
49	1202	12	100.0	99	1603	16	>100.0
50	901	9	>100.0				

*As displayed in Figure 2.2 and defined in Appendix B.
Source: NRDI Computer Outputs. 1975.

flow conditions, and the basins are ranked from lowest to highest
BOD dilution concentration (Table 6.1). Similarly, for the SS
dilution ranking, note the concentrations of SS in milligrams per
liter are listed for low and average flow conditions, and basins are
ranked from lowest to highest SS dilution concentration (Table 6.2).

Comparison of Water Quality Data

Characterizing the quality of the nation's water is difficult
for several reasons (McKelvey 1974). Among them is the fact that
there is no universal set of criteria that can be used as a guide to
describe natural water conditions; also, the lack of extensive baseline data makes it difficult to make reasonable projections of future
levels of water quality characteristics.

The U.S. Geological Survey, the Council of Environmental
Quality, EPA, and other groups are currently involved in efforts
to expand the amount of baseline data available to make projections
of levels of future water quality. These new data should improve
descriptions of current conditions and make possible more accurate
projections of water quality characteristics.

The NCWQ environmental contractors conducted water quality
analyses and environmental impact assessments at 41 selected sites
throughout the nation. A majority of these sites are segments of
various rivers throughout the coterminous United States, and the
remaining portion includes coastal areas and estuaries and several
"standing water" sites, including lakes, reservoirs, and impoundments.

NRDI simulations estimate BOD concentrations for the 99
basins. Some of the NCWQ site contractors also estimated BOD
concentrations. Both data sets show BOD concentrations for 1973
conditions and BPT/ST and BAT/BPWTT objectives. The NRDI
concentrations show averages for entire basins, whereas those
estimated by the NCWQ are representative of specific sites. Thus,
the concentrations estimated were not expected to be identical;
however, a positive correlation was anticipated between these two
data sets. A reasonable expectation would be that when one data
set indicated relatively degraded water quality in a river reach or
basin, the other would also; and when one data set projected improvements in water quality in river setments or basins, the other would
project similar improvements.

To verify the NRDI water quality index, BOD concentrations
generated by NRDI simulations were compared with NCWQ contractor analyses. NCWQ estimates of BOD (summer condition) and

TABLE 6.3

NRDI and NCWQ BOD Concentration Estimates for Selected Basins

(milligrams per liter)

Waterbody	NRDI Basin Code	1973 Conditions		BPT/ST		BAT/BPWTT	
		NRDI	NCWQ	NRDI	NCWQ	NRDI	NCWQ
Biscayne Bay	0305	7.2	1.1	3.0	0.6	1.3	0.6
Boston Harbor	0103	9.5	6.3	2.2	6.3	1.2	5.0
Connecticut River	0105	6.9	2.8	1.1	2.0	0.7	1.7
Guadelupe River	1205	2.5	2.3	1.3	2.2	0.4	1.2
Housatonic River	0104	3.5	6.3	0.7	2.0	0.5	1.3
Hudson River	0201	4.2	2.0	1.0	1.0	0.7	1.0
Iowa–Cedas River	0703	1.6	4.1	0.8	2.4	0.4	2.3
Lower Columbia River	1702	0.1	1.7	0.0	1.1	0.0	1.0
Lower Missouri River	1011	0.8	3.2	0.3	3.1	0.3	2.6
Potomac River	0206	11.9	1.4	3.5	0.7	1.6	0.7
St. Johns River	0101	0.6	3.2	0.1	3.1	0.1	2.3
Snake River	1703	0.8	1.4	0.6	1.3	0.2	1.3
South Platte River	1007	41.5	5.7	41.5	4.3	15.9	3.0
Susquehanna River	0204	13.5	3.4	NA*	NA	NA	NA
Trinity River	1202	16.4	12.6	8.5	2.0	2.1	2.0
Upper Rio Grande River	1302	49.9	3.9	49.9	3.9	17.5	2.7

*Data not available.

Source: NRDI Computer Outputs. 1976.

NRDI estimates (low flow conditions) were compared for a sample of 16 water bodies and the corresponding NRDI river basins (see Table 6.3). The sample represents approximately 80 percent of the rivers assessed by the NCWQ and those that had BOD projections for areas covered by the NRDI.

A comparison of NCWQ and NRDI estimates of BOD concentrations by categorizing each area as degraded or not degraded showed that 50 percent of the basins were positively correlated. The comparison defined "degraded" conditions as a BOD concentration greater than 2.5 milligrams per liter for 1973 conditions, greater than 2.0 for BPT/ST conditions, and greater than 1.5 for BAT/BPWTT conditions. Specifically, both data sets indicated that seven basins were relatively degraded (high concentrations) and that one basin was nondegraded (low concentrations) under 1973 conditions. For the other eight the two data sets indicated different conditions. Either the NCWQ data indicated that water quality was relatively degraded and NRDI indicated it was relatively nondegraded, or vice versa.

After applying BPT/ST, the NCWQ and NRDI projections of BOD concentrations are as well correlated as they were in 1973. The NCWQ and NRDI projections were positively correlated on seven of the basins and negatively correlated on the other nine basins. In the latter case, the NCWQ more often reported a high BOD concentration, whereas the NRDI reported a low BOD concentration.

After applying BAT/BPWTT, the NCWQ and NRDI projections of BOD concentrations were more positively correlated than under either 1973 or BPT/ST conditions. The NCWQ and NRDI projections were positively correlated for ten basins and negatively correlated for the other six basins. In this case, none of the NCWQ projections indicated a low BOD concentration, whereas the NRDI indicated a high concentration.

Another comparison between the two sets of data was made to see whether they indicated the same trend in the percentage of the basins that changed in relative water quality after applying BPT/ST and BAT/BPWTT. For 1973 conditions, NCWQ data show that 47 percent of the sample has BOD concentrations ranging from 0.0 to 2.9 milligrams per liter compared to 33 percent for the same interval as estimated by NRDI.

After applying BPT/ST, the NCWQ estimates show 73 percent of the sample with BOD concentrations between 0.0 and 2.9 milligrams per liter compared to 67 percent for the NRDI. The NCWQ data show a 36 percent increase in the number of river basins with BOD ranging from 0.0 to 2.9 milligrams per liter compared to a 51 percent increase as suggested by NRDI data.

After applying BAT/BPWTT, both NCWQ and NRDI data show that 87 percent of the sample is expected to have BOD concentrations ranging from 0.0 to 2.9 milligrams per liter. This represents a 16 percent increase for NCWQ estimates and a 23 percent increase for NRDI estimates. However, NCWQ data show larger percentages of the sample with BOD ranging from 0.0 to 2.9 milligrams per liter at 1973 and BPT/ST (1977) conditions as compared to NRDI estimates. By 1983, the NCWQ estimates that 100 percent of the sample would achieve a BOD concentration of at least 5.9 milligrams per liter, whereas the NRDI estimates that only 87 percent of the sample would achieve this level, with the remaining percent having BOD greater than or equal to 12 milligrams per liter.

NRDI estimates of BOD for the South Platte and Upper Rio Grande basins must be viewed skeptically because their flows are very low compared to other basins within the same region. For the South Platte basin (region 10) and the Upper Rio Grande basin (region 13), NRDI low flow estimates are approximately 1 and 4 percent, respectively, of the average low flows for all basins in those regions. In addition, 85 percent of the BOD generated in the South Platte and 98 percent of the BOD in the Upper Rio Grande comes from municipal sources. These discharges combined with the low flows described above explain the high BOD concentrations in the areas after application of BAT/BPWTT.

In summary, the NRDI water quality ranking index for selected basins is neither verified nor refuted by comparing it to the NCWQ data. The NRDI data indicate relative water quality rankings similar to NCWQ data for approximately one-half the sites. Neither NRDI nor NCWQ estimates of 1973 conditions compare very well with EPA data for a more limited set of rivers. Both NRDI and NCWQ projections of BOD indicate the same general trend in improvements in water quality. After meeting the BPT/ST objective, the NCWQ projects that 73 percent of the rivers would have BOD concentrations ranging from 0.0 to 2.9 milligrams per liter, whereas the NRDI projects that 67 percent would have concentrations in the same category. After meeting the BAT/BPWTT objectives, both the NCWQ and NRDI project that 87 percent of the rivers would have BOD concentrations in the least degraded category.

REFERENCES

McKelvey, V. E. 1974. "Water Quality—Is It Getting Better or Worse?" Paper presented at the 7th International Water Quality Symposium at Washington, D. C., mimeographed.

National Commission on Water Quality. 1976. Staff Report to the
    National Commission on Water Quality. Washington, D.C.:
    Government Printing Office.
U.S. Department of Commerce. 1975. List of Coastal Counties.
    Office of Coastal Zone Management, mimeographed.
Wollman, Nathaniel, and Gilbert Bonem. 1971. Outlook for Water—
    Quality, Quantity, and National Growth. Baltimore: Johns
    Hopkins Press.

7

MEETING THE 1977 AND 1983 OBJECTIVES OF PUBLIC LAW 92-500

This chapter summarizes relevant data from the preceding chapters, integrating estimates of residual generation and discharge, and reduction technology costs for municipal, industrial, and areal-source activities. Further, this chapter provides a comprehensive assessment of the BPT/ST and BAT/BPWTT policies required by PL 92-500.

NATIONAL TOTALS

Table 7.1 shows that nationally, municipal activities generate 62 percent of the point-source BOD but only 1 percent of point-source SS. Industrial activities generate 38 percent of the point-source BOD and 99 percent of point-source SS. Mining activities alone account for 96 percent of the industrial generation of SS. Table 7.2 shows the capital investments required to meet 1973, BPT/ST, and BAT/BPWTT conditions.

Industrial activities had reduced their BOD discharge by 46 percent and SS discharge by 37 percent, with a capital investment of $6.1 billion by 1973, while municipal activities had achieved reductions of 56 percent in BOD and 58 percent in SS, with a much larger capital investment of $32.5 billion. Consequently, the relative share of BOD for point sources of 1973 discharge shows industrial activities accounting for 43 percent of the total and municipal for 57 percent. Their relative shares of SS were not altered significantly by investment made by 1973.

The projected residual discharges and costs of residual reduction technology with the installation of BPT/ST show a

TABLE 7.1

1973 BOD and SS Generation and Discharge by Control Alternative
(billion pounds per year)

Source	BOD				SS			
	1973 Generation	1973 Discharge	BPT/ST Discharge	BAT/BPWTT Discharge	1973 Generation	1973 Discharge	BPT/ST Discharge	BAT/BPWTT Discharge
Point sources (PS)	21.4	10.3	2.7	1.3	1,162.6	729.0	14.4	11.9
percent of national total	69	51	22	12	45	34	1	1
Industrial	8.2	4.5	0.9	0.4	1,148.5	723.0	12.7	11.1
percent of PS	38	44	33	31	99	99	88	93
Municipal	13.2	5.8	1.8	0.9	14.1	6.0	1.7	0.8
percent of PS	62	56	67	69	1	1	12	7
Areal sources (AS)	9.8	9.8	9.8	9.8	1,430	1,430	1,430	1,430
percent of national total	31	49	78	88	55	66	99	99
Urban Runoff	3.3	3.3	3.3	3.3	59.9	59.9	59.9	59.9
percent of AS	34	34	34	34	4	4	4	4
Non-irrigated agriculture	6.5	6.5	6.5	6.5	1,370	1,370	1,370	1,370
percent of AS	66	66	66	66	96	96	96	96
National totals	31.2	20.1	12.5	11.1	2,600	2,160	1,440	1,440

Source: NRDI Computer Outputs. 1975.

TABLE 7.2

Costs of Meeting 1973, BPT/ST and BAT/BPWTT Objectives

Activity	Replacement Value of Existing (1973) Waste Treatment Facilities (billions of dollars)	Percent of Total	Capital Costs of Additional Facilities to Meet BPT Objectives (billions of dollars)	Percent of Total	Capital Costs of Additional Facilities to Meet BAT/BPWTT Objectives (billions of dollars)	Percent of Total
Industries						
Pulp and paper	1.9		1.8		0.6	
Organic chemicals	0.3		1.3		1.3	
Petroleum refining	0.3		0.6		1.2	
Iron and steel	0.8		1.9		1.3	
Inorganic chemicals	0.3		0.5		0.1	
Plastics and synthetics	0.4		0.3		0.2	
Textiles	0.2		0.6		0.3	
Steam/electric power	0.0		1.7		0.0	
Mining	0.0		0.3		0.0	
All other	1.9		2.7		1.8	
Industry total	6.1	15	11.7	29	6.8	40
Municipalities	32.5	85	28.4	61	10.3	60
Total*	38.6	100	40.1	100	17.1	100

*Applies to the point-source category only.

Source: NRDI Computer Outputs, 1975.

considerable difference between industrial and municipal activities.
Industrial activities are estimated to have invested $11.6 billion
to limit their residual discharge to .9 billion pounds of BOD and
12.6 billion pounds of SS—80 and 98 percent, respectively, of their
1973 discharge. Municipal activities are estimated to have invested
$28.4 billion to limit their residual discharge to 1.8 billion pounds
of BOD and 1.7 billion pounds of SS—70 and 72 percent, respectively,
of their 1973 discharge. The more significant reduction achieved by
industrial activities means that industrial activities after installing
BPT would account for only 33 percent of point-source BOD and 88
percent of point-source SS.

Projected residual discharges and costs after installation of
BAT/BPWTT are similar. Industrial activities are estimated to
have invested $7.0 billion to limit their residual discharge to .4
billion pounds of BOD and .9 billion pounds of SS—55 and 1 percent,
respectively, of their discharge after application of BPT. Municipal
activities are estimated to have invested $10.3 billion to limit their
residual discharge to .9 billion pounds of BOD and .8 billion pounds
of SS—50 and 53 percent, respectively, of their discharge after
application of ST. After the achievement of BAT/BPWTT, municipal
sources would account for approximately twice as much point-source
BOD as industrial activities and only one-sixth of the point-source
SS as industrial activities. Both types of activities would be more
comparable in SS discharge if the SS discharge of 10.0 billion pounds
from mining activities were excluded from the industrial total.

While the installation of BPT for industrial activities would
result in a considerable reduction in residual discharge in proportion to the capital investment, the installation of BAT would not
accomplish the same degree of reduction for a comparable cost
(Tables 7.1 and 7.2).

The capital cost of industrial residuals reduction technology
would increase markedly from 1973 conditions after the installation
of BPT and BAT. Capital costs are estimated to increase from an
in-place replacement value of $6.4 billion to $18.1 billion, or by
180 percent after the installation of BPT, and to $24.9 billion, or
by 290 percent, with installation of BAT. The additional expenditure
of $6.8 billion, or 58 percent of the BPT costs, to install BAT would
decrease BOD discharge by an additional 12 percent and SS by an
additional 1 percent over the reduction achieved between 1973 and
BPT levels.

Similarly, the installation of ST technology for municipal
activities would result in a considerable reduction in residual discharge in proportion to the capital investment, but the installation
of BPWTT would not accomplish the same degree of reduction for

a comparable cost. The capital cost of municipal residuals reduction in technology would increase markedly from 1973 conditions
after the application of ST and BPWTT. Capital costs are estimated
to increase from an in-place replacement value of $35.5 billion to
$66.1 billion, or by 86 percent, after the installation of ST tech-
nology and to $77.1 billion or by 117 percent with installation of
BAT. The additional expenditure of $11.0 billion, or 36 percent of
the ST technology cost, to install BPWTT would decrease BOD
discharge by an additional 16 percent and SS by an additional 15
percent over the reduction achieved between 1973 and ST technology
levels.

A comparison of industrial and municipal residuals reductions
and the costs of applying BPT and ST reveals that residual reduction in the industrial sector is a more cost-effective expenditure
(see Table 7.3). Municipal BOD discharge is estimated to decrease
from the 1973 discharge by only 4.0 billion pounds after a capital
expenditure of $28.4 billion for ST technology. Industrial discharge
of BOD would be reduced by slightly less than the municipal reduction for a capital expenditure of slightly more than one-third. Thus,
the ratio of BOD residual removal to costs is 293 for industrial
activities compared to 141 for municipal activities. Similarly, the
ratio of SS residual removal to costs is 61,241 for industrial activities compared to 151 for municipal activities.

Another interesting comparison is of the relative residual
reductions and costs among industrial activities (Table 7.3). First,
the ratio of residual removed to costs reveals that BPT removals
of BOD and SS for all industries are more cost effective than BAT
removals of the same residuals. The only exception is SS removal
for inorganic chemicals. The fact that more residuals are removed
in proportion to dollar expenditure with BPT than with BAT cor-
responds to the data presented in Table 7.1. Second, the usual
case is for some industries to have more cost effective removal
of BOD and for other industries to have more cost effective removal
of SS. On the one hand, the organic chemicals, pulp and paper, and
plastics and synthetics industries have the highest removal efficiencies for BOD. On the other hand, the mining, steam/electric power,
and iron and steel industries have the highest removal efficiencies
of SS. If additional data were displayed for other residuals, such
as COD and heavy metals, they would reveal yet other patterns of
cost effectiveness. This variability in cost effectiveness patterns
suggests that no one industry be automatically excluded from
effluent limitations because of a low cost effectiveness removal
on one parameter since it would probably fall into the high cost
effectiveness category for another parameter. Third, some industries have relatively higher cost effectiveness removal rates for

TABLE 7.3

Industrial and Municipal Residuals Reduction and Costs of Meeting BPT/ST and BAT/BPWTT Objectives, for 1973 Conditions

BOD

Activity	BOD after Application of BPT/ST				BOD after Application for BAT/BPWTT			
	Residual Discharge (millions of pounds)	Residual Removed (millions of pounds)	Capital Cost of Technology (billions of dollars)	Ratio of Residuals Removed to Cost (pounds per year per thousand dollars)	Residual Discharge (millions of pounds)	Residual Removed (millions of pounds)	Capital Cost of Technology (billions of dollars)	Ratio of Residuals Removed to Cost (pounds per year per thousand dollars)
Industrial								
Pulp and paper	370	1,529	1.8	849	185	185	0.6	308
Organic chemicals	173	1,061	1.3	816	36	137	1.3	105
Petroleum refining	11	103	0.6	62	4	7	1.2	6
Iron and steel	6	33	1.9	17	1	5	1.3	4
Inorganic chemicals	4	1	0.5	2	3	3	0.1	30
Plastics and synthetics	30	173	0.3	576	6	24	0.2	120
Textiles	26	100	0.6	167	8	18	0.3	60
Steam/electric power	0	0	1.7	NA	0	0	0.0	NA
Mining	6	5	0.3	17	6	0	0.0	0
All other	300	586	2.7	217	168	132	1.8	73
All industry	930	3,570	11.7	305	407	500	6.8	74
Municipal	1,800	4,000	28.4	141	900	900	10.3	87

SS

Activity	SS after Application of BPT/ST				SS after Application of BAT/BPWTT			
	Residual Discharge (millions of pounds)	Residual Removed (millions of pounds)	Capital Cost of Technology (billions of dollars)	Ratio of Residuals Removed to Cost (pounds per year per thousand dollars)	Residual Discharge (millions of pounds)	Residual Removed (millions of pounds)	Capital Cost of Technology (billions of dollars)	Ratio of Residuals Removed to Cost (pounds per year per thousand dollars)
Industrial								
Pulp and paper	390	1,667	1.8	1,665	146	244	0.6	407
Organic chemicals	0	0	1.3	0	0	0	1.3	0
Petroleum refining	1	28	0.6	47	0.5	0.9	1.2	0.8
Iron and steel	814	1,735	1.9	1,733	68	746	1.3	574
Inorganic chemicals	691	454	0.5	908	345	346	0.1	3,460
Plastics and synthetics	25	91	0.3	303	2	23	0.2	115
Textiles	18	92	0.6	153	4	14	0.3	47
Steam/electric power	817	5,060	1.7	2,976	817	0	0.0	0
Mining	9,617	700,404	0.3	2,334,680	9,617	0	0.0	0
All other	242	464	2.7	172	75	167	1.8	93
All industry	12,700	710,300	11.7	60,709	11,100	1,600	6.8	235
Municipal	1,700	4,300	28.4	151	800	900	10.3	87

Source: NRDI Computer Outputs, 1975.

both BOD and SS. The pulp and paper, inorganic chemicals, and plastics and synthetics industries fall into this category. This pattern indicates that, on a national basis, more residual reduction would occur for a limited investment by requiring a few industries to achieve higher removal efficiencies rather than having all industries achieve uniformly lower removal efficiencies.

The cost effectiveness analysis indicates significant differences within the industrial sector and between the municipal and industrial sectors. Given a limited commitment of financial resources for achieving water quality, more reduction would occur toward the 1977 goal with emphasis on the industrial sector. However, comparable residual reduction with a limited budget would occur toward the 1983 goal, which means that emphasis on either source would achieve similar results.

The 1972 Act focuses on controlling residuals from municipal and industrial activities. However, there are other activities that discharge BOD and SS, and other residuals from point and areal sources, which affect water quality. The following discussion first presents a reasonably comprehensive view of BOD and SS residuals from four major activities and then describes the magnitude of N and P residuals from municipal and nonirrigated agricultural activities.

Table 7.1 shows that, in 1973, point sources were clearly the dominant source of BOD residuals generation, accounting for approximately 70 percent of the total. Within their respective categories, municipal acitvities and nonirrigated agriculture were the most significant sources. However, point sources accounted for slightly less SS (45 percent) than areal sources. Within their respective categories, industrial activities and nonirrigated agriculture were clearly the dominant sources.

The 1973 residual reduction technology in place for point sources reduced the potential discharge to an actual discharge of 20 billion pounds of BOD and 2,160 billion pounds of SS. Point sources accounted for approximately the same BOD as areal sources, while for SS, point sources were clearly less important than areal sources.

After the application of BPT/ST, point sources of BOD would account for only 22 percent of the national BOD total and areal sources, primarily nonirrigated agriculture, would account for the other 78 percent. Point sources of SS would account for only 1 percent of the national total and areal sources, almost exclusively nonirrigated agriculture, would account for the other 99 percent.

After the installation of BAT/BPWTT, total discharge would be decreased to 11 and 4 percent of 1973 BOD and SS, respectively.

Point sources would then account for 12 percent of total BOD and
still only 1 percent of the total SS. Thus, areal sources, assuming
no control of these activities during the ten-year period, would
clearly dominate the national total of BOD and SS discharge; whether
areal sources would be dominant in all areas of the country is discussed later in this chapter.

The other types of residuals reasonably well covered by the
NRDI are P and N. The coverage is only partial because there was
inadequate data for N and P discharges from industrial and urban
runoff activities. Consequently, the discussion is limited to the
two major sources, municipal and nonirrigated agriculture activities.

Residual generation in 1973 was estimated to be 5.8 billion
pounds per year of N and 1.9 billion pounds per year of P (see
Table 7.4). Municipal activities accounted for 2.5 billion pounds
of N, or 43 percent of the national total, and for .6 billion pounds
of P, or 32 percent of the national total.

TABLE 7.4

Generation and Discharge of Nitrogen and
Phosphorus from Selected Sources
(billion pounds per year)

Residual	Municipal Sources	Non-irrigated Agriculture	Total
Nitrogen			
1973 generation	2.5	3.3	5.8
1973 discharge	2.0	3.3	5.3
After application of BPT/ST	1.8	3.3	5.1
After application of BAT/BPWTT*	1.8	1.7	3.5
Phosphorus			
1973 generation	0.6	1.3	1.9
1973 discharge	0.6	1.3	1.9
After application of BPT/ST	0.4	1.3	1.9
After application of BAT/BPWTT*	0.3	0.7	1.0

*With agriculture controlled.
Source: NRDI Computer Outputs. 1975.

The 1973 residual reduction technology in place for municipal activities did not appreciably affect the relative importance of activities as sources of P and N residuals.

Application of ST and BPWTT will increase the demand for nonirrigated agriculture as a source of P and N. Thus, after installation of BPWTT, the municipal share would drop to 35 and 19 percent of N and P discharges, respectively (assuming no control of nonirrigated agriculture). If both BPWTT and agricultural controls were implemented the municipal share would be 51 and 30 percent of N and P discharges, respectively.

In summary, the application of BPT/ST and BAT/BPWTT would significantly reduce BOD and SS residuals from point-source activities, but would leave other significant sources unaffected. The control of point sources would decrease their 1973 level of BOD and SS by 87 and 98 percent, respectively. However, areal sources would then discharge approximately 8 times more BOD than point sources and 100 times more SS.

Similarly, the application of BPT/ST and BAT/BPWTT would significantly reduce P and N residuals from these activities, but would leave unaffected other significant sources of the same residuals. Nonirrigated agricultural activities would account for three times more N than municipal sources and five times more P.

While the 1972 Law's emphasis on controlling point sources does leave significant uncontrolled residuals from areal sources, residuals from these sources may not have as significant an impact on water quality as residuals from point sources. However, areal sources of nutrient residuals could be as significant as residuals from point sources in that they accumulate in receiving waters and would be available to stimulate algal growth.

REGIONAL ANALYSIS

The spatial distribution of both the costs and changes in residual discharges involved in meeting PL 92-500 technology objectives is of prime importance. Showing the diversity of both costs and changes in residual discharges is important because incurring more or less cost does not, in itself, indicate more or less change.

A regional analysis also assists in evaluating the influences of activities not subject to technological controls by the law. Although PL 92-500 does not mandate controls for nonpoint sources, they are nevertheless large and in some regions they are the dominant discharger. Thus, a regional evaluation including all

TABLE 7.5

BOD and SS Discharges and Capital Costs of Control Technology,
for 1973, BPT/ST, and BAT/BPWTT Conditions, by Region

Water Resource Region	Technology Objective	Point-Source BOD (millions of pounds)			Point-Source SS (millions of pounds)			Capital Costs (millions of dollars)		
		Industrial	Municipal	Total	Industrial	Municipal	Total	Industrial	Municipal	Total
New England	1973	150	340	490	1,900	330	2,230	220	810	1,030
	BPT/ST	30	70	100	160	60	220	660	1,150	1,810
	BAT/BPWTT	10	60	70	90	50	140	220	200	220
Middle Atlantic	1973	570	1,260	1,830	9,770	1,280	11,050	790	3,230	4,020
	BPT/ST	110	330	440	700	330	1,041	1,560	5,280	6,840
	BAT/BPWTT	50	200	250	450	180	630	820	1,180	2,000
South Atlantic	1973	640	610	1,250	3,170	670	3,840	1,130	2,110	3,240
	BPT/ST	180	170	350	460	150	610	1,350	3,160	4,510
	BAT/BPWTT	80	90	170	230	80	310	760	950	1,710
Great Lakes	1973	420	610	1,030	27,590	690	28,280	890	3,500	4,390
	BPT/ST	80	290	370	2,540	310	2,850	1,900	3,560	5,460
	BAT/BPWTT	40	90	130	2,110	80	2,190	980	1,360	2,340
Ohio	1973	310	450	750	4,740	490	5,230	760	2,280	3,040
	BPT/ST	70	140	210	620	140	750	1,600	2,280	3,040
	BAT/BPWTT	30	60	90	220	50	280	940	1,870	2,810
Tennessee	1973	160	90	250	9,530	100	9,630	170	300	470
	BPT/ST	30	20	50	200	20	220	260	360	620
	BAT/BPWTT	10	10	20	170	10	180	150	80	230
Upper Mississippi	1973	270	300	570	7,100	300	7,400	300	1,970	2,270
	BPT/ST	60	120	180	1,080	120	1,200	620	2,340	2,960
	BAT/BPWTT	30	70	100	990	70	1,060	280	480	760
Lower Mississippi	1973	310	80	390	1,500	110	1,610	380	580	960
	BPT/ST	60	40	100	190	40	230	610	750	1,360
	BAT/BPWTT	30	30	60	120	30	150	530	340	870
Souris–Red–Rainy	1973	10	10	20	120	10	130	20	40	60
	BPT/ST	0	0	0	10	0	10	20	70	90
	BAT/BPWTT	0	0	0	0	0	0	10	0	10

Region		C1	C2	C3	C4	C5	C6	C7	C8	C9
Missouri	1973	40	250	290	392,170	250	392,420	160	1,320	1,480
	BPT/ST	20	90	110	3,250	90	3,340	280	1,670	1,950
	BAT/BPWTT	10	80	90	3,220	80	3,300	190	100	290
Arkansas–White–Red	1973	130	120	250	5,610	120	5,730	230	910	1,140
	BPT/ST	30	50	80	150	50	200	420	1,370	1,790
	BAT/BPWTT	20	10	30	110	10	120	280	1,100	1,380
Texas Gulf	1973	620	160	780	2,570	190	2,760	470	1,180	1,650
	BPT/ST	100	90	190	170	90	260	1,070	1,660	2,730
	BAT/BPWTT	30	10	40	90	10	100	870	1,820	2,690
Rio Grande	1973	20	10	30	16,490	10	16,500	20	220	240
	BPT/ST	0	10	10	330	10	340	110	330	330
	BAT/BPWTT	0	0	0	230	0	230	40	200	240
Upper Colorado	1973	0	10	10	3,790	10	3,800	0	30	30
	BPT/ST	0	0	0	170	0	170	50	50	100
	BAT/BPWTT	0	0	0	110	0	110	10	10	20
Lower Colorado	1973	10	60	70	94,620	80	94,700	30	120	150
	BPT/ST	0	10	10	770	10	780	70	240	310
	BAT/BPWTT	0	10	10	770	10	780	20	10	30
Great Basin	1973	10	20	30	86,830	30	86,860	40	160	200
	BPT/ST	0	10	10	880	10	890	90	210	300
	BAT/BPWTT	0	0	0	810	0	810	60	110	170
Pacific Northwest	1973	470	100	570	39,550	100	39,650	470	740	1,210
	BPT/ST	80	40	120	480	40	520	320	680	1,000
	BAT/BPWTT	40	40	80	420	40	460	220	80	300
California	1973	210	1,260	1,470	15,530	1,290	16,820	340	1,920	2,260
	BPT/ST	50	270	320	440	260	700	670	2,300	2,970
	BAT/BPWTT	20	140	160	350	110	460	460	410	870
National totals	1973	4,350	5,750	10,100	722,580	6,050	728,630	6,100	32,500	38,600
	BPT/ST	900	1,760	2,660	12,590	1,720	14,310	11,600	28,400	40,000
	BAT/BPWTT	400	900	1,300	12,540	820	13,360	6,800	10,300	17,100

Source: NRDI Computer Outputs. 1975.

97

sources provides a more accurate indication of the real changes in
residual discharges that can be expected by meeting technology
objectives for point sources. Alternatives to uniformity based upon
regional diversity are presented in Chapter 8.

The data present two levels of geographic disaggregation,
the 18 regions that reveal the major geographic differences in the
relative magnitude of residual discharges and technology costs for
major activities, and the 99 river basins. Basins analysis yields
the same general conclusions as the regional analysis, indicating
that the regional analysis does not completely obscure most local
differences.

1973 Conditions

The magnitude of residual discharges and the replacement
value of the in-place residual reduction technology varied considerably in 1973 among the 18 regions (see Table 7.5). In seven
of these regions, industrial activities contributed a majority of
the total point-source BOD discharge. Two East Coast regions
(Middle and South Atlantic), one Midwest region (Great Lakes),
and one Far West region (California) accounted for approximately
80 percent of the total point-source BOD discharge. Industrial
discharges are twice as great as municipal discharges in the
Lower Mississippi, Texas Gulf, Rio Grande, and Pacific Northwest regions and are less than one-half of municipal in the New
England, Middle Atlantic, Missouri, Upper Colorado and Lower
Colorado, and California regions.

Three western regions (Missouri, Lower Colorado and Great
Basin) accounted for approximately 80 percent of the total point-
source SS discharges. In all regions, industrial SS discharges are
significantly greater than municipal discharges.

In all regions the replacement value of municipal control
technology exceeds the replacement value of industrial control
technology.

Per capita residuals discharges and control costs are shown
in Table 7.6. This view gives a different perspective on diversity
of residuals and costs due to the large differences in population
among regions. Per capita BOD discharges in 1973 averaged 48
pounds while ranging from 15 to 81 pounds. The three regions with
the highest per capita BOD discharges were the Pacific Northwest
(81 pounds), Texas Gulf (78 pounds), and California (70 pounds).
The large industrial contribution compared to municipal in the first
two regions explains their high per capita figure; the greater use
of municipal sewerage treatment plants is a possible explanation

of the high per capita figure in California. Per capita SS discharges in 1973 averaged 3,440 pounds while ranging from 150 to 47,350 pounds. The three regions with significantly higher per capita BOD discharges were the Great Basin (86,860 pounds), Lower Colorado (47,350 pounds), and Missouri (43,600 pounds). These high per capita discharges were attributable to mining activities. Per capita replacement values of residual reduction technology averaged $182 and ranged from $75 to $200.

BPT/ST Objectives

The projected discharges and costs of residual reduction with the application of BPT/ST show considerable variation among regions. In all regions, municipal ST costs are greater than the industrial BPT costs. The costs of residual reduction in four regions account for 60 percent of the total costs of BPT/ST. After this investment, industrial activities would account for a majority of the total point-source BOD discharge in only five regions while municipal activities are dominant in the remaining regions. However, this capital investment would not alter the 1973 condition in which industrial activities accounted for 51 percent or more of the total point-source discharges of SS in all regions.

Per capita BOD discharges after meeting BPT/ST objectives would average 13 pounds and range from 0 to 19 pounds. Per capita SS residual discharges after meeting BPT/ST objectives would average 76 pounds and would range from 17 to 890 pounds. Per capita costs for meeting the BPT/ST objectives would average $189 and range from $140 to $333.

BAT/BPWTT Objectives

The projected residuals discharges and costs of residuals reduction with installation of BAT/BPWTT also show considerable regional variation. In this case, industrial costs to achieve BAT are greater than the municipal costs to achieve BPWTT in eight of the regions. The notable exceptions, that is, where the municipal investment is twice as great as the industrial investment, are the Arkansas-White-Red, Texas Gulf, and Rio Grande regions. Again the costs of residual reduction are concentrated in four regions, which account for 60 percent of the projected BAT/BPWTT costs. After this investment, industrial activities would account for a majority of the total point-source BOD discharge in four regions. However, this investment does not alter the BPT/ST condition

TABLE 7.6

Per Capita Point-Source Residuals Discharge and Control Costs, by Region

Water Resource Region	Technology Objective	BOD Per Capita (pounds)	SS Per Capita (pounds)	Capital Costs Per Capita (dollars)
New England	1973	41	185	$ 86
	BPT/ST	8	18	150
	BAT/BPWTT	6	12	18
Middle Atlantic	1973	46	276	100
	BPT/ST	11	24	171
	BAT/BPWTT	6	12	50
South Atlantic	1973	50	153	130
	BPT/ST	14	31	310
	BAT/BPWTT	7	16	68
Great Lakes	1973	34	942	146
	BPT/ST	12	95	182
	BAT/BPWTT	4	73	78
Ohio	1973	35	249	144
	BPT/ST	10	35	237
	BAT/BPWTT	4	13	133
Tennessee	1973	63	2,408	118
	BPT/ST	13	55	155
	BAT/BPWTT	5	45	58
Upper Mississippi	1973	44	569	175
	BPT/ST	14	92	228
	BAT/BPWTT	8	82	58
Lower Mississippi	1973	65	28	160
	BPT/ST	17	38	227
	BAT/BPWTT	10	25	145
Souris-Red-Rainy	1973	33	216	100
	BPT/ST	0	17	150
	BAT/BPWTT	0	0	16

Missouri	1973	32	43,602	164
	BPT/ST	12	371	217
	BAT/BPWTT	10	367	32
Arkansas–White–Red	1973	36	819	163
	BPT/ST	11	29	256
	BAT/BPWTT	4	17	197
Texas Gulf	1973	78	276	165
	BPT/ST	19	26	273
	BAT/BPWTT	4	10	269
Rio Grande	1973	15	8,250	120
	BPT/ST	5	170	165
	BAT/BPWTT	0	115	120
Upper Colorado	1973	0	12,633	100
	BPT/ST	0	566	333
	BAT/BPWTT	0	366	66
Lower Colorado	1973	35	47,350	75
	BPT/ST	0	390	155
	BAT/BPWTT	0	39	15
Great Basin	1973	30	86,860	200
	BPT/ST	10	890	300
	BAT/BPWTT	0	810	170
Pacific Northwest	1973	81	5,664	173
	BPT/ST	17	74	143
	BAT/BPWTT	11	66	43
California	1973	70	801	108
	BPT/ST	15	33	141
	BAT/BPWTT	8	22	41
National average	1973	48	3,436	$182
	BPT/ST	13	67	189
	BAT/BPWTT	6	63	81

Source: NRDI Computer Outputs. 1975.

TABLE 7.7

Comparison of BOD and SS Discharges From Point and Areal Sources (millions of pounds)

Region	Technology Objective	BOD Total Point	BOD % of Area Total	BOD Urban Runoff	BOD % of Area Total	BOD Nonirrigated Agriculture	BOD % of Area Total	BOD Total, All Sources	SS Total Point	SS % of Area Total	SS Urban Runoff	SS % of Area Total	SS Nonirrigated Agriculture	SS % of Area Total	SS Total, All Sources
New England	1973	500	57	330	37	50	6	880	2,220	16	6,400	47	4,930	37	13,550
	BPT/ST	90	19	330	69	50	12	470	220	2	6,400	55	4,930	43	11,550
	BAT/BPWTT	70	15	330	73	50	12	450	140	1	6,400	56	4,930	43	11,470
Middle Atlantic	1973	1,830	71	690	27	80	2	2,600	11,050	32	11,720	34	11,810	34	34,580
	BPT/ST	440	37	690	57	80	6	1,210	1,010	4	11,720	48	11,810	48	24,540
	BAT/BPWTT	250	25	690	68	80	7	1,020	620	3	11,720	49	11,810	49	24,150
South Atlantic	1973	1,250	65	570	30	100	5	1,920	3,850	9	11,130	27	26,270	64	41,250
	BPT/ST	340	34	570	57	100	10	1,010	610	2	11,130	29	26,270	69	38,010
	BAT/BPWTT	170	20	570	68	100	12	840	310	1	11,130	30	26,270	70	37,710
Great Lakes	1973	1,040	58	620	34	140	8	1,800	28,280	50	9,810	17	18,720	33	56,810
	BPT/ST	370	33	620	55	140	12	1,130	2,850	9	9,810	31	18,720	60	31,380
	BAT/BPWTT	130	15	620	70	140	16	890	2,190	7	9,810	32	18,720	61	30,720
Ohio	1973	750	41	310	17	750	41	1,810	5,230	3	5,260	3	151,930	94	162,420
	BPT/ST	210	17	310	25	750	58	1,270	750	—	5,260	3	151,930	96	157,940
	BAT/BPWTT	90	8	310	27	750	65	1,150	280	—	5,260	3	151,930	96	157,470
Tennessee	1973	250	48	50	10	220	42	520	9,630	19	1,000	2	40,980	79	51,610
	BPT/ST	50	16	50	16	220	68	320	820	2	1,000	2	40,980	95	42,800
	BAT/BPWTT	20	7	50	17	220	76	290	180	—	1,000	2	40,980	97	42,160
Upper Mississippi	1973	570	34	110	7	970	59	1,650	7,400	4	2,160	1	175,360	95	184,920
	BPT/ST	180	14	110	9	970	77	1,260	1,190	1	2,160	1	175,360	98	178,710
	BAT/BPWTT	100	8	110	9	970	83	1,180	1,060	1	2,160	1	175,360	98	178,580
Lower Mississippi	1973	390	37	110	10	550	53	1,050	1,610	1	2,230	1	182,940	98	186,780
	BPT/ST	110	14	110	14	550	72	770	240	—	2,230	1	182,940	99	185,410
	BAT/BPWTT	60	8	110	15	550	77	720	150	—	2,230	1	182,940	99	185,320
Sourie-Red-Rainy	1973	20	40			30	60	50	10	—			3,550	100	3,560
	BPT/ST					30	100	30	10	—			3,550	100	3,560
	BAT/BPWTT					30	100	30	10	—			3,550	100	3,560
Missouri	1973	290	14	50	3	1,650	84	1,990	392,420	55	970	—	325,060	45	718,450
	BPT/ST	110	6	50	3	1,650	91	1,810	3,340	1	970	—	325,060	99	329,370
	BAT/BPWTT	90	5	50	3	1,650	92	1,790	3,300	1	970	—	325,060	99	329,330
Arkansas-White-Red	1973	250	45	50	10	250	45	550	5,740	7	890	1	78,980	92	85,610
	BPT/ST	80	21	50	13	250	66	380	200	—	890	1	78,980	99	80,070
	BAT/BPWTT	30	9	50	16	250	75	330	120	—	890	1	78,980	99	79,990
Texas Gulf	1973	780	77	180	18	50	5	1,010	2,760	9	3,480	11	25,520	80	31,760
	BPT/ST	190	45	180	43	50	12	420	260	1	3,480	12	25,520	87	29,260
	BAT/BPWTT	50	18	180	64	50	18	280	100	—	3,480	12	25,520	88	29,100
Rio Grande	1973	30	75			10	25	40	16,500	75			5,470	25	21,970
	BPT/ST	10	50			10	50	20	340	6			5,470	94	5,810
	BAT/BPWTT					10	100	10	230	4			5,470	96	5,700
Upper Colorado	1973	10	1			720	99	730	3,800	3			127,530	97	131,330
	BPT/ST					720	100	720	180	—			127,530	100	127,710
	BAT/BPWTT					720	100	720	110	—			127,530	100	127,640
Lower Colorado	1973	80	57			60	43	140	94,690	79			25,010	21	119,700
	BPT/ST	20	25			60	75	80	790	3			25,010	97	25,800
	BAT/BPWTT	10	14			60	86	70	730	3			25,010	97	25,790
Great Basin	1973	30	50	10	17	20	33	60	86,850	93	210	—	5,840	6	92,900
	BPT/ST	10	25	10	25	20	50	40	890	13	210	3	5,840	84	6,940
	BAT/BPWTT			10	33	20	67	30	810	12	210	3	5,840	85	6,860
Pacific Northwest	1973	570	54	110	10	380	36	1,060	39,650	34	1,880	2	75,910	65	117,440
	BPT/ST	130	21	110	18	380	61	620	520	1	1,880	2	75,910	97	78,310
	BAT/BPWTT	80	14	110	20	380	67	570	460	1	1,880	2	75,910	97	78,250
California	1973	1,470	69	140	7	530	25	2,140	16,820	16	2,750	3	87,900	82	107,560
	BPT/ST	320	32	140	14	530	54	990	700	1	2,750	3	87,900	96	91,360
	BAT/BPWTT	160	19	140	17	530	64	830	680	1	2,750	3	87,900	96	91,390
National totals	1973	10,100	51	3,320	17	6,530	33	19,950	728,000	34	59,900	3	1,373,784	63	2,162,180
	BPT/ST	2,700	22	3,320	26	6,530	52	12,550	14,300	1	59,900	4	1,373,784	95	1,447,898
	BAT/BPWTT	1,900	12	3,320	30	6,530	58	11,150	13,400	1	59,000	4	1,373,784	95	1,444,901

Source: NRDI Computer Outputs, 1975.

in which industrial activities are the dominant source of SS in all regions.

Per capita BOD residual discharges after meeting BAT/BPWTT objectives would average 6 pounds and range from 0 to 11. Per capita SS residual discharges after meeting BAT/BPWTT objectives would average 81 pounds and range from 0 to 810 pounds. Per capita costs for meeting the BAT/BPWTT objectives would average $81 and range from $15 to $269.

Other Sources of BOD and SS

In 1973 point sources were significant dischargers of BOD and SS residuals (see Table 7.7). Point sources of BOD were greater dischargers than urban runoff sources in all regions, and municipal sources alone were greater dischargers than urban runoff sources in 14 out of the 18 regions. Even in the four regions where urban runoff sources were greater dischargers than municipal sources, the differences between the two categories were so small that estimation techniques rather than actual conditions could account for the dominance of urban runoff sources. Point sources of BOD were greater dischargers than nonirrigated agricultural sources in 13 regions. Only in the Midwest and West were nonirrigated agricultural sources more significant dischargers of BOD. Point sources of SS were greater dischargers than urban runoff sources in 12 of the regions, but municipal sources alone were greater dischargers only in those four regions with insignificant urban runoff activities. Point sources of SS were greater dischargers than nonirrigated agricultural sources in only five regions, primarily because of mining activities in the West. Nonirrigated agricultural sources were greater dischargers than point sources in the other 13 regions.

After the application of BPT/ST, point-source BOD would no longer be the dominant source in many regions. Urban runoff sources of BOD would be greater dischargers than point sources in eight regions, primarily in the East and Midwest, and would be greater dischargers than municipal sources in ten regions. The relative importance of nonirrigated agricultural BOD would also increase as these sources dominate point sources in 13 regions. Urban runoff SS would be greater than point-source SS in 12 regions. The relative significance of nonirrigated agricultural sources of SS compared to point sources would also shift because they would dominate point sources in all 18 regions.

After the installation of BAT/BPWTT, point-source BOD would be less significant than after BPT/ST. Urban runoff sources of BOD would then be greater dischargers than point sources in 12

TABLE 7.8

Number of Basins with Varying Ranges of BOD Concentration,
by Region and Technology Objective

Region and Technology Objective	BOD Concentration under Low Flow Conditions					BOD Concentration under Average Flow Conditions				
	0-2.9	3.0-5.9	6.0-8.9	9.0-11.9	>12	0-2.9	3.0-5.9	6.0-8.9	9.0-11.9	>12
New England										
1973	3	1	1	1	0	4	1	0	1	0
BAT/ST	6	0	0	0	0	5	0	1	0	0
BPT/BPWTT	6	0	0	0	0	5	0	1	0	0
Middle Atlantic										
1973	1	2	1	1	1	3	3	0	0	0
BAT/ST	5	1	0	0	0	6	0	0	0	0
BPT/BPWTT	6	0	0	0	0	6	0	0	0	0
South Atlantic										
1973	6	2	1	0	0	8	1	0	0	0
BAT/ST	8	1	0	0	0	9	0	0	0	0
BPT/BPWTT	9	0	0	0	0	9	0	0	0	0
Great Lakes										
1973	4	1	0	1	2	4	2	1	0	1
BAT/ST	6	1	0	0	1	6	1	0	0	1
BPT/BPWTT	7	0	1	0	0	6	1	0	0	1
Ohio										
1973	2	3	1	0	1	5	1	1	0	1
BAT/ST	6	1	0	0	0	6	1	0	0	0
BPT/BPWTT	7	0	0	0	0	6	1	0	0	0
Tennessee										
1973	1	1	0	0	0	1	1	0	0	0

BAT/ST	2	0	0	0	0	2	0	0	0	0	0
BPT/BPWTT	2	0	0	0	0	2	0	0	0	0	0
Upper Mississippi											
1973	4	1	0	0	0	4	1	0	0	0	0
BAT/ST	5	0	0	0	0	4	1	0	0	0	0
BPT/BPWTT	5	0	0	0	0	5	0	0	0	0	0
Lower Mississippi											
1973	3	0	0	0	0	3	0	0	0	0	0
BAT/ST	3	0	0	0	0	3	0	0	0	0	0
BPT/BPWTT	3	0	0	0	0	3	0	0	0	0	0
Souris–Red–Rainy											
1973	0	0	1	0	0	1	0	0	0	0	0
BAT/ST	1	0	0	0	0	1	0	0	0	0	0
BPT/BPWTT	1	0	0	0	0	1	0	0	0	0	0
Missouri											
1973	8	1	0	1	2	5	1	1	2	2	2
BAT/ST	10	0	0	0	1	5	2	0	2	2	2
BPT/BPWTT	10	0	0	0	1	5	2	0	2	2	2
Arkansas–White–Red											
1973	2	3	0	1	1	3	2	1	0	1	1
BAT/ST	6	0	0	0	1	3	3	0	0	1	1
BPT/BPWTT	6	0	1	0	0	5	1	0	0	1	1
Texas Gulf											
1973	2	0	0	0	3	1	2	1	0	1	1
BAT/ST	2	1	2	0	0	2	1	0	0	0	0
BPT/BPWTT	5	0	0	0	0	4	1	0	0	0	0
Rio Grande											
1973	3	0	1	0	1	2	1	0	2	0	0
BAT/ST	3	1	0	0	1	2	1	2	0	0	0
BPT/BPWTT	3	1	0	0	1	2	2	1	0	0	0

(continued)

(Table 7.8 continued)

Region and Technology	BOD Concentration under Low Flow Conditions					BOD Concentration under Average Flow Conditions				
Objective	0-2.9	3.0-5.9	6.0-8.9	9.0-11.9	>12	0-2.9	3.0-5.9	6.0-8.9	9.0-11.0	>12
Upper Colorado										
1973	3	0	0	0	0	0	0	0	2	1
BAT/ST	3	0	0	0	0	0	0	0	2	1
BPT/BPWTT	3	0	0	0	0	0	0	0	2	1
Lower Colorado										
1973	3	0	0	0	0	0	0	0	2	1
BAT/ST	3	0	0	0	0	0	0	0	2	1
BPT/BPWTT	3	0	0	0	0	0	0	0	2	1
Great Basin										
1973	0	1	0	0	3	0	0	0	0	4
BAT/ST	1	1	0	0	3	0	0	0	0	4
BPT/BPWTT	1	0	1	1	1	0	0	0	0	4
Pacific Northwest										
1973	7	0	0	0	0	6	0	0	0	1
BAT/ST	7	0	0	0	0	6	0	0	0	1
BPT/BPWTT	7	0	0	0	0	6	0	0	0	1
California										
1973	2	0	1	0	4	3	2	0	0	2
BAT/ST	3	0	2	0	2	4	1	0	1	1
BPT/BPWTT	3	0	2	0	2	4	1	1	0	1
National Totals										
1973	54	16	7	4	18	53	18	5	9	14
BAT/ST	80	6	4	0	9	64	13	3	7	12
BPT/BPWTT	87	1	5	1	5	69	9	3	6	12

Note: Low flow conditions include only point-source discharges, while average flow conditions include point- and nonpoint-source discharges. Concentrations are in milligrams per liter.

Source: NRDI Computer Outputs, 1975.

regions; nonirrigated agricultural BOD would then be greater than point-source BOD in 14 regions. However, after the installation of BAT/BPWTT, point-source SS would decrease only slightly, thus not changing the pattern of BPT/ST.

Water Quality Impacts

Under the 1973 residual discharge conditions, the NRDI water quality index suggests that water quality varied considerably among and within regions (see Table 7.8). Only in three regions were all basins in the least degraded class. Among all 99 basins, 52 were in the least degraded category. Most of the regions had at least half of their basins in the two least degraded classes, despite the existence of several basins within those regions in which the water quality might be severely degraded by the 1973 level of discharge.

Regulation of point sources to achieve BPT/ST would have a noticeable impact upon water quality in most regions as measured by the water quality index. In 13 regions, over 70 percent of the basins within the region would achieve water quality within the least degraded class, and 79 basins among all the regions would fall into the same class. In 16 regions, over 70 percent of the basins within the region, and 89 basins among all the regions, would achieve a value within the three best classes. However, there would still remain several basins scattered throughout the country in which the dilution index remains very high.

Regulation of point sources to achieve BAT/BPWTT would have only a limited impact upon water quality in most regions as measured by the water quality index. Eighty-five basins would fall into the best class (rather than the 79 that would with BPT/ST). Ninety-two basins among all the regions would achieve water quality within the three classes that indicate the least degradation. A significant reduction in the number of basins in the classes indicating severe degradation of water quality is found in three regions. However, there would remain basins in several regions that would still have severe water quality degradation even after achievement of BAT/BPWTT.

Similar conclusions about 1973, BPT/ST, and BAT/BPWTT conditions emerge from comparing the water quality index based on average flow, which includes the impact of areal sources. The major difference is the impact of areal sources on water quality in midwestern and western regions after the achievement of BPT/ST and BAT/BPWTT. Areal sources, apparently nonirrigated agriculture, would prevent the improvement of all waters in those regions.

FIGURE 7.1

Basins in which Municipal BOD Discharge is Greater Than Industrial BOD Discharge

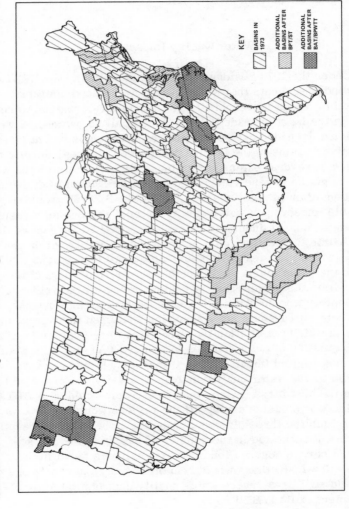

Source: <u>NRDI Computer Outputs.</u> 1975.

RIVER BASIN DATA

At the river basin level of aggregation, data on the costs of residual reduction technology for point sources and the relative magnitude of residual discharges from major activities are best displayed graphically rather than in tabular form. Basin mapping indicates the geographic location of residual discharges with sufficient specificity to be meaningful for policy analysis.

The magnitude of residuals discharge varied considerably in 1973 among the 99 basins. Total BOD discharge from point sources of over 1.5 million pounds per day are found in the basins containing New York and Los Angeles, not surprising given the population in these two regions. Total point-source discharges of over one million pounds per day are found in one basin in the Middle Atlantic region and in the Dallas-Houston basin. Other basins with large point-source discharges are most often those with large populations.

The relative contributions of municipal and industrial activities to the point-source discharge total in 1973 varied considerably. Municipal activities in 59 basins accounted for a majority of the total point-source BOD discharged, and in no basins accounted for a majority of the total point-source SS discharged.

A geographical comparison of municipal and industrial sources illustrates the particular basins where municipal activities were the dominant source of BOD in 1973 (see Figure 7.1). BOD discharge from municipal activities is the major point-source discharge in over half of the basins including those around New York City, parts of the Appalachians, parts of the Great Lakes region, the Missouri River basin, the West, and California. This dominance in the New York region can be explained by its population and in the Great Lakes by large cities that have only primary treatment of municipal sewage. In the Missouri River basin and the West, the dominance can be explained by the absence of industrial activities.

After the installation of BPT/ST, the relative contribution of residuals by municipal and industrial activities would not change significantly from 1973 conditions. Municipal activities would dominate in BOD discharge in approximately 10 percent more basins (see Table 7.9).

After the installation of BAT/BPWTT, the relative contribution of residuals by municipal and industrial activities would change less than between 1973 and application of BAT/ST. Municipal activities would dominate in BOD discharge in 72 basins and in SS in 11 basins.

WATER POLLUTION CONTROL
                          TABLE 7.9

                Relative Dominance of Residual Discharge for
                       Municipal and Industrial Sources

| | | Point-Source Discharge | | | | | |
| | Control | (in percent) due to Municipal Sources | | | | | |
Residual	Alternative	0-14	15-29	30-49	50-69	70-84	85-100
BOD	1973	10	13	13	21	19	23
	BPT/ST	1	11	17	23	25	22
	BAT/BPWTT	3	12	21	24	24	24
SS	1973	73	21	3	2	0	0
	BPT/ST	39	23	29	7	1	0
	BAT/BPWTT	45	20	22	7	4	0

Source: NRDI Computer Outputs. 1976.

Municipal activities accounted for more of the technology replacement value in 90 basins and industrial activities dominated in the remaining 9. Even if the replacement value of interceptors is excluded from the municipal total, municipal activities would dominate in 60 basins.

After installation of BPT/ST, costs to achieve ST are greater than industrial BPT costs in 60 basins. After the installation of BAT/BPWTT, costs to achieve BPWTT are greater than industrial BAT costs in 58 basins.

In 1973, point sources of BOD dominated urban runoff sources in 98 basins (Table 7.10); municipal sources alone dominated urban runoff in 82 basins. The exceptions, where urban runoff BOD dominated point-source BOD, were in Florida (see Figure 7.2). Similarly, point sources dominated urban runoff in SS discharge in 75 basins, and municipal alone dominated urban runoff in 50 basins.

After the installation of BPT/ST, urban runoff dominates point-source discharges of BOD and SS in 15 and 20 basins, respectively.

After the installation of BAT/BPWTT, urban runoff dominates point-source BOD and SS in 29 and 44 basins, respectively. The significance of this fact is enhanced since only 44 basins have any urban runoff residuals.

In 1973, nonirrigated agricultural sources of BOD dominated point sources in 68 basins; (Table 7.11); point sources dominated nonirrigated agricultural sources only in New England basins and

FIGURE 7.2

Basins in Which Urban Runoff BOD Discharge is Greater Than Point-Source BOD Discharge

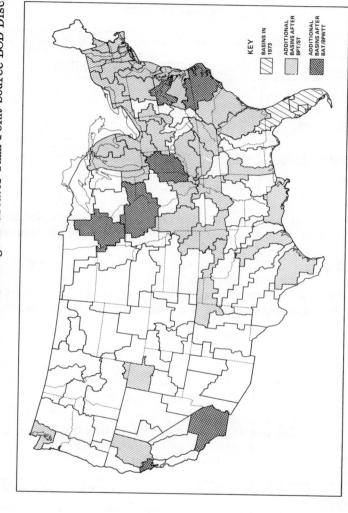

Source: NRDI Computer Outputs. 1975.

TABLE 7.10

Relative Dominance of BOD Discharge for
Urban Runoff and Point Sources
(number of basins)

Control Alternative	Urban Runoff and Point Sources due to Urban Runoff (in percent)					
	0–14	15–29	30–49	50–69	70–84	85–100
1973	58	17	22	2	0	0
BPT/ST	54	2	8	27	7	1
BAT/BPWTT	54	0	2	14	17	12

Source: NRDI Computer Outputs. 1976.

the West. Nonirrigated agricultural sources of SS dominated point
sources of SS in 80 basins. Geographic patterns of this dominance
are shown in Figure 7.3.

After BPT/ST, nonirrigated agricultural sources dominate
point sources of BOD and SS in 70 and 85 basins, respectively.

After BAT/BPWTT, nonirrigated agricultural sources would
dominate point sources of BOD and SS in 80 and 90 basins, respectively.

TABLE 7.11

Relative Dominance of BOD Discharge for
Agricultural and Point Sources
(number of basins)

Control Alternative	Agricultural and Point Source due to Agriculture (in percent)					
	0–14	15–29	30–49	50–69	70–85	85–100
1973	38	15	10	15	7	14
BPT/ST	23	7	15	15	13	26
BAT/BPWTT	12	11	10	12	18	36

Source: NRDI Computer Outputs. 1975.

FIGURE 7.3

Basins in Which Nonirrigated Agricultural BOD Discharge is Greater Than Point-Source BOD Discharge

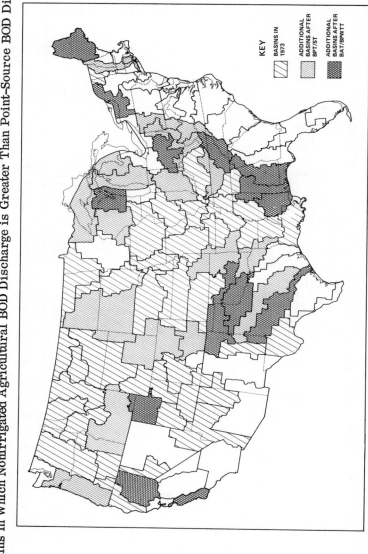

KEY

BASINS IN
1973

ADDITIONAL
BASINS AFTER
BPT/ST

ADDITIONAL
BASINS AFTER
BAT/BPWTT

Source: NRDI Computer Outputs. 1975.

TABLE 7.12

BOD Discharge Based on 1973 and Future Production and Population
(millions of pounds per year)

Residuals Source Category	1973, In Place	1973, BPT/ST	1977, BPT/ST	Projection Year and Abatement Policy 1983, BPT/ST	1973, BAT/BPWTT	1983, BAT/BPWTT
Pulp and paper	1,899	370	390	432	185	247
Organic chemicals	1,233	173	176	186	37	51
Petroleum refining	114	11	11	12	4	6
Iron and steel	39	6	6	7	1	2
Inorganic chemicals	5	4	4	5	3	4
Plastics and synthetics	203	30	31	32	6	8
Textiles	126	26	26	28	8	10
Steam/electric power (no growth)	0	0	0	0	0	0
Mining	11	6	6	6	6	6
Other industry	886	300	315	359	168	227
Industry subtotal	4,500	930	960	1,070	420	560
Municipal subtotal	5,800	1,800	2,000	2,400	900	1,400
Point-source total	10,300	2,700	3,000	3,500	1,300	2,000

Note: Industry data pertain to BPT and BAT; municipal data, to ST and BPWTT.
Source: NRDI Computer Outputs. 1975.

114

EFFECTS OF INDUSTRY AND POPULATION GROWTH
ON RESIDUALS DISCHARGE

The evaluation of the uniform BPT/ST and BAT/BPWTT policies so far presented has been based on 1972-73 industrial production and population. However, these policies would actually be effective at much later dates, presumably 1977 and 1983. Industrial production and population at those later dates would affect residual discharges and costs.

In order to assess the effect of industry and population growth, a separate analysis was performed based on the Wharton two-digit SIC industrial production projection and the census series E population projection. The annual growth in each industry is assumed to be discharging residuals at the NSPS level, which is approximated by the application of BAT. The growth projection is simply the percentage increase in production over the specified time period. This type of projection assumes that industrial output would be increased by building a mix of new plants comparable to the mix existing in 1973. This assumption may overestimate residual discharge because the raw waste load from new plants may be less than older plants, and may overestimate costs since it does not allow for increased production at existing facilities. For municipal sources, the future-year population is assumed to be discharging residuals at the ST or BPWTT level specified for the facilities modeled for the year 1990.

Comparison of BOD discharge after BPT/ST and after BAT/BPWTT, based on 1973 production and population and on 1977 or 1985 production and population, shows only a small difference in residual discharges (see Table 7.12).

Industrial BOD discharge after BPT is estimated at .9 billion pounds per year based on 1973 production levels, and 10 percent less than 1977 levels. If BPT rather than NSPS levels were applied to the annual growth, then industrial BOD discharge would be 1.0 billion pounds per year. Industrial discharge after BAT is estimated at .4 billion pounds per year based on 1973 production and .6 billion pounds per year based on 1983 production. If only BPT rather than NSPS levels were applied to the annual growth, then industrial discharge would be 1.2 billion pounds per year. The municipal increase shown in Table 7.10 is greater than the industrial increase because the existing population does not have to meet uniformly higher treatment levels.

The most interesting difference in estimated residual discharge is between the 1977 projection, after application of BPT and the 1983 projection, after BPT for industrial dischargers. The 1983 estimate

TABLE 7.13

Capital Costs Based on 1973 and Future Production and Population

(billions of dollars)

Residuals Source Category	Projection Year and Abatement Policy					
	1973, In Place	1973, BPT/ST	1977, BPT/ST	1983, BPT/ST	1973, BAT/BPWTT	1983, BAT/BPWTT
Pulp and paper	1.9	1.8	2.1	3.1	0.6	1.6
Organic chemicals	0.3	1.3	1.6	2.4	1.3	2.1
Petroleum refining	0.3	0.6	0.6	1.1	1.2	1.7
Iron and steel	0.8	1.9	1.8	4.8	1.3	4.2
Inorganic chemicals	0.3	0.5	0.5	0.8	0.1	0.2
Plastics and synthetics	0.4	0.3	0.3	0.6	0.2	0.4
Textiles	0.2	0.6	0.6	1.2	0.3	0.7
Steam/electric power	0.0	1.7	2.3	3.1	0.0	0.0
Mining	0.0	0.3	0.3	0.3	0.0	0.0
Other industry	1.9	2.7	3.2	4.8	1.8	3.3
Industry subtotal	6.1	11.7	13.3	22.2	6.8	14.2
Municipal subtotal	32.5	28.4	30.7	38.1	10.3	14.3
Point-source total	38.6	40.0	43.8	60.0	17.3	28.3

Note: Industry data pertain to BPT and BAT; municipal data, to ST and BPWTT.

Source: NRDI Computer Outputs. 1975.

describes the condition in which production in 1973 has only to install BPT in 1983 and in which new production has to meet BAT objectives. The level of industrial BOD discharge would be approximately 10 percent greater in 1983 than in 1977 due to the increase in production and a decision not to require more stringent technology (BAT) for existing (1973) industrial activities. Only with the installation of BAT for 1973 production would industrial residual discharge be lower in 1983 than in 1977. A similar comparison for municipalities cannot be made because there is no new source performance standard for municipal activities.

Capital costs to meet BPT/ST and BAT/BPWTT objectives based on 1973 production are somewhat less than the technology costs to serve 1977 and 1983 production and population (see Table 7.13). BPT is estimated to cost $11.7 billion based on 1973 production levels, but the additional cost of NSPS technology to handle the increase in production between 1973 and 1977 would add $1.6 billion, to give a total technology cost of $13.3 billion. If the BPT goal is not realized until 1983, then the additional costs of NSPS technology to handle the increase in production between 1973 and 1983 would add $10.5 billion, to give total BPT/NSPS costs of $22.2 billion.

If only BPT rather than NSPS technology is required for new production, then the additional cost to handle the increase in production would add $1.3 billion, to change the total 1977 technology cost to $13.0 billion. If only BPT rather than NSPS technology is required for all new production between 1973 and 1983, then the additional costs would be $7.6 billion, to give total BPT/NSPS costs of $19.3 billion. BAT is estimated to cost $6.8 billion based on 1973 production, but the additional costs of NSPS technology to handle the increase in production between 1977 and 1983 would add $7.4 billion, to give total costs of $14.2 billion.

For municipalities, ST technology is estimated to cost $28.4 billion based on 1973 population, but is estimated to cost $30.7 billion to provide the same level of treatment for the 1977 population. BPWTT is estimated to cost $10.3 billion based on 1973 population, but would actually cost $14.3 billion to provide the same level of treatment for the 1983 population.

In summary, changing the production and population basis for estimating residuals and costs does not significantly after most of these estimates. The BPT-level residual estimate based on 1973 production is only 4 percent less than it would be if based on 1977 production and the BAT-level residual estimate based on 1973 production is 25 percent less than it would be if based on 1983 production. In either case, the 1983 discharge would only be approximately

10 percent of the 1973 discharge. The ST-level residual estimate based on 1973 population is only 12 percent less than it would be if based on 1977 population and the BAT-level residual estimate based on 1973 production is 36 percent less than it would be if based on 1983 population. In either of these cases, the 1983 discharge would only be approximately 20 percent of the 1973 discharge. Similarly, the BPT cost estimate based on 1973 production is only 11 percent less than it would be if based on 1977 production, and the BAT cost estimate based on 1973 production is 50 percent less than it would be if based on 1983 production. While a 50 percent difference is significant, this difference is exaggerated by assuming that all new growth occurs in new plants. The ST cost estimate based on 1973 population is 7 percent less than it would be if based on 1977 population, and the BPWTT cost estimate is 28 percent less than it would be if based on 1983 population.

REFERENCE

National Research Council. 1975. National Residuals Discharge Inventory Computer Outputs. Washington, D.C.: National Research Council.

8

ALTERNATIVES TO
THE UNIFORM APPLICATION
OF BAT/BPWTT OBJECTIVES

Chapter 7 described the changes in residual discharges, relative water quality, and control costs that would be associated with achieving the BPT/ST and BAT/BPWTT objectives of the 1972 Act. The results show that a uniform application of BPT/ST objectives could achieve significant reduction in discharges of residuals and improvement in one measure of water quality, while application of more stringent BAT/BPWTT objectives could result in relatively little change in either residual discharges or relative water quality despite a large additional capital investment. In light of these findings, four alternative policies to the uniform application of the more stringent BAT/BPWTT objectives have been examined.

Alternative policy 1 requires that BAT/BPWTT be applied only in those basins that have relatively severe water quality problems.

Alternative policy 2 considers all sources of residuals, point and nonpoint, in a basin and selects a cost-effective strategy for achieving the same levels of residuals removal as achieved by the uniform application of BAT/BPWTT.

Alternative policy 3 requires that only new industrial growth meet BAT/BPWTT objectives, thereby limiting 1973 production levels to meeting only BPT objectives.

Alternative policy 4 does not require meeting BAT/BPWTT objectives in some or all counties that have potential for ocean discharge.

The analysis of these alternatives adequately illustrates costs and consequences at the national level, but should not be taken to reflect unique situations (for example, river segments).

ALTERNATIVE POLICY 1: LIMITING BAT/BPWTT OBJECTIVES TO AREAS WITH RELATIVELY POOR WATER QUALITY

River basins with similar residuals discharges but different surface water flows will experience different resultant water quality with the uniform application of abatement technologies. Alternative policy 1 is to attempt to take into account the dilution effects of surface waters on the quantities of residuals discharges; using assimilative capacity essentially substitutes natural for technological processes for neutralizing the harmful effects of residuals.

The approach chosen to illustrate the alternative limits the application of BAT/BPWTT to basins that have a BOD dilutions index greater than or equal to 3.0 milligrams per liter during low flow conditions (those basins that are not ranked in the highest water quality category as described in Chapter 6). Only BPT/ST is applied in all other basins. Low flow conditions are used because the dilution index (water quality) under low flow is more sensitive to changes in point-source discharges than under average flow conditions, when large loadings from nonpoint sources tend to dominate the effects of changes in point-source loadings on water quality.

If uniform water quality is a policy objective, then it could be achieved at significantly less cost—25 to 30 percent of uniform BAT/BPWTT—by applying BAT/BPWTT to only 21 of the 99 basins (see Table 8.1). In the remaining 78 basins, BAT/BPWTT may not really be necessary, since they may have adequate water quality after BPT/ST. For example, in the Upper Mississippi region all five basins are in the highest water quality category after BPT/ST. However, in 14 basins, BAT/BPWTT may be inadequate to achieve sufficiently good water quality. For example, in the Great Basin region, even after BAT/BPWTT, only one out of four basins is the highest water quality category.

The value of 3.0 milligrams per liter was arbitrarily defined for demonstration purposes as "good" water quality since neither a national modeling capability nor a universal definition of good water quality exists to permit more precision. The numerical value of the quantitative results is of less importance than the insights gained. One insight is that regional diversity of ambient water should be an important factor in determining the benefits of uniform BAT/BPWTT objectives. A second is that a selective application of BAT/BPWTT may have the same consequences for water quality as the uniform application, but with significantly lower costs.

TABLE 8.1

Effects of Alternative Policy 1

Region	Uniform BAT/BPWTT		Alternative Policy 1			Alternative as a Percentage of Uniform BAT/BPWTT	
	BOD Discharge (millions of lbs/yr)	Capital Cost (millions of $)	Basins Affected*	BOD Discharge (millions of lbs/yr)	Capital Cost (millions of $)	BOD	Capital Cost
New England	67	415	0	94	0	140	0
Mid Atlantic	250	1,979	1	414	258	166	13
South Atlantic	172	1,694	1	336	72	195	4
Great Lakes	125	2,315	2	201	1,049	161	45
Ohio	88	2,778	1	169	445	192	16
Tennessee	24	230	0	49	0	204	0
Upper Mississippi	97	754	0	177	0	182	0
Lower Mississippi	58	857	0	110	0	190	0
Souris-Red-Rainy	4	9	0	5	0	125	0
Missouri	94	248	1	108	53	115	21
Arkansas-White-Red	28	1,368	1	79	47	282	3
Texas Gulf	45	2,659	3	73	1,809	162	68
Rio Grande	5	236	2	10	82	200	35
Upper Colorado	1	15	0	2	0	200	0
Lower Colorado	14	36	2	16	20	114	56
Great Basin	4	186	3	5	156	125	84
Pacific Northwest	81	297	0	128	0	158	0
California	165	857	4	173	757	105	88
National Totals (rounded)	1,300	17,100	21	2,100	4,700	161	27

*If 0, Alternative Policy 1 requires no additional treatment.

Source: NRDI Computer Outputs. 1975.

121

ALTERNATIVE POLICY 2: COST-EFFECTIVE OPTIONS TO
UNIFORM BAT/BPWTT OBJECTIVES

Other sources of residuals, areal activities, have not been
considered in uniform application of BAT/BPWTT. Areal activities,
if controlled, could result in the same quantity of residuals removal
in an area as with the uniform application of BPT/ST and/or BAT/
BPWTT at less cost. Such a nonuniform policy would permit a BAT
level of control in one industry before BPT in another, if cost-
effective.

This alternative addresses the issue of cost-effective options
by considering other sources (nonirrigated agriculture and urban
runoff) of residuals, and applying BPT/BPWTT nonuniformly, in
a cost-effective manner to all sources. To illustrate this alternative, cost-effective options have been formulated for each basin to
achieve the BOD removals that would be obtained from uniform
application of BPT/ST and BAT/BPWTT to point sources. This
alternative assumes that the detrimental effect of residuals from
all sources is similar and that the only decision criterion is cost
minimization. Possible differential effects by source are not
considered.

Cost-effective options were performed as follows. Table 8.2
is a sample of the NRDI outputs discussed below. First, the amount
of BOD removed per year with the application of BPT/ST and
BAT/BPWTT was computed for each point-source category in
each basin. Similarly, the amount removed when nonirrigated
agriculture and urban runoff are controlled is also calculated based
on the technologies described in Chapter 5. Second, the capital
cost at each control level (BPT/ST and BAT/BPWTT for point
sources and one level for each areal-source category) is divided
by the amount of BOD removed at each control level in each source
category. These dollar-per-pound-of-removal values are then used
as measures of the cost effectiveness of BOD removal for each
category in each basin (despite the fact that some of the costs are
attributable to the reduction in discharge of other residuals).
Finally each basin is then analyzed as follows (see Table 8.3). The
amounts of BOD removed by the uniform application of BPT/ST and
BAT/BPWTT in each basin are considered the target removal values
for the basin. A cost-effective option is created by sequentially
selecting categories with the lowest cost-effective index (dollars
per pound value), regardless of whether they are point or areal
sources, until reaching the BPT/ST and then the BAT/BPWTT
target removal quantities. While the total cost of BPT/ST in the

TABLE 8.2

Selected NRDI Outputs for the White-Patoka, Wabash Basin

Activity Category	Bod Removal (million pounds per year)				Unit Capital Costs (dollars per pound)		
	Current	BPT/ST	BAT/BPWTT	Areal Source Control	BPT/ST	BAT/BPWTT	Areal Sources
Pulp and paper	5.23	3.64	0.49	—	2.30	4.59	—
Organic chemicals	1.56	2.73	0.35	—	1.77	10.00	—
Petroleum refining	2.61	0.25	0.09	—	17.20	165.00	—
Inorganic chemicals	0.00	0.00	0.00	—	0.00	0.00	—
Plastics and synthetics	0.22	0.38	0.05	—	1.36	8.32	—
Textiles	0.00	0.00	0.00	—	0.00	0.00	—
Mining	0.00	0.07	0.00	—	27.80	0.00	—
Other industry	19.60	8.76	2.28	—	2.03	7.45	—
Municipal	133.00	24.00	17.00	—	9.38	8.73	—
Urban runoff	—	—	—	5.78	—	—	60.10
Agricultural	—	—	—	36.00	—	—	3.76
Total	162.00	39.80	21.00	41.80	—	—	—

Source: NRDI Computer Outputs. 1976.

123

TABLE 8.3

Effects of Alternative Policy 2

Region	Uniform Policy[a]		Alternative Policy 2[b]		Alternative as a Percentage of Uniform BAT/BPWTT	
	BOD Removed (millions of lbs/yr)	Capital Cost (millions of $)	BOD Removed (millions of lbs/yr)	Capital Cost (millions of $)	BOD	Capital Cost
New England	428	2,207	428	1,760	100	80
Mid Atlantic	1,575	8,747	1,575	7,794	100	89
South Atlantic	1,081	6,156	1,081	5,237	100	85
Great Lakes	911	7,713	911	6,439	100	83
Ohio	666	7,703	666	3,629	100	47
Tennessee	224	837	224	423	100	51
Upper Mississippi	474	3,680	474	888	100	24
Lower Mississippi	336	2,205	336	277	100	13
Souris–Red–Rainy	20	92	20	86	100	93
Missouri	192	2,183	192	327	100	15
Arkansas–White–Red	223	3,132	223	1,275	100	41
Texas Gulf	740	5,360	740	4,815	100	90
Rio Grande	24	557	24	404	100	73
Upper Colorado	8	121	8	2	100	2
Lower Colorado	63	342	63	166	100	49
Great Basin	22	483	22	229	100	47
Pacific Northwest	494	1,279	494	494	100	39
California	1,302	3,794	1,302	3,204	100	84
National Totals (rounded)	8,800	57,200	8,800	37,500	100	66

[a]Uniform BPT/ST followed by Uniform BAT/BPWTT.
[b]Cost-effective BPT/ST followed by Cost-effective BAT/BPWTT.
Source: NRDI Computer Outputs, 1975.

TABLE 8.4

Effects of Alternative Policy 3*

Region	Uniform BAT		Alternative Policy 3		Alternative as a Percentage of Uniform BAT	
	BOD Discharge (millions of lbs/yr)	Capital Cost (millions of $)	BOD Discharge (millions of lbs/yr)	Capital Cost (millions of $)	BOD	Capital Cost
New England	16	593	30	374	188	63
Mid Atlantic	71	1,824	130	999	183	55
South Atlantic	109	1,729	204	968	187	56
Great Lakes	53	2,657	96	1,671	181	63
Ohio	38	2,564	79	1,627	208	63
Tennessee	15	330	34	177	227	54
Upper Mississippi	39	631	70	353	179	56
Lower Mississippi	34	937	73	411	215	44
Souris–Red–Rainy	1	14	2	8	200	57
Missouri	12	357	19	164	158	46
Arkansas–White–Red	21	549	36	267	171	49
Texas Gulf	44	1,596	113	726	257	45
Rio Grande	3	78	5	42	167	54
Upper Colorado	0	24	0	17	0	71
Lower Colorado	2	57	3	35	150	61
Great Basin	1	142	2	86	200	61
Pacific Northwest	55	701	97	246	176	35
California	31	947	52	440	168	46
National Totals (rounded)	500	14,500	1,000	8,600	200	59

*BOD and costs are shown for industry only for 1983 activities.

Source: NRDI Computer Outputs, 1975.

125

selected basin was $290 million, the alternative which achieves
the same removal costs only $123 million. Similarly, BAT/BPWTT
requires $200 million compared to $125 million for the cost-
effective alternative.

Table 8.4 displays the national capital cost savings that can
be obtained by substituting a cost-effective approach for uniform
requirements. It shows that if the water quality, using residuals
removed as a proxy measure, is a policy objective, then it can be
achieved at 30 to 35 percent less total cost. The results show that
the amount of BOD removed throughout the nation after uniform
BPT can be removed at a 60 percent reduction in cost by following
a cost-effective approach.

The analysis also shows that controlling BOD discharge from
nonirrigated agriculture was very cost effective. Some level of
control was cost effective in 91 basins. However, urban runoff
control proved to be cost effective in only two basins, not surprising given the control scenario used. If additional BOD removal
technology had been included in the control scenario, it is likely
that urban runoff would have proven cost effective in a greater
number of basins.

Although only an illustration, the analysis clearly suggests
that substantial savings can be produced and the same residuals
removals achieved by pursuing a cost-effective approach rather
than uniform BPT/ST and BAT/BPWTT requirements.

ALTERNATIVE POLICY 3: REQUIRING ONLY NEW PRODUCTION
TO MEET MORE STRINGENT TREATMENT LEVELS

Some proponents of more stringent uniform treatment requirements justify BAT/BPWTT as necessary to offset the effects of
increasing industrial production and population growth, which generate additional residuals. They argue that the gains achieved with
BPT/ST would be lost without the application of more stringent
technology. However, an alternative level between BPT/ST and
BAT/BPWTT is a requirement that new industrial production meet
NSPS and that 1973 industrial production meet only BPT. NSPS
are generally comparable to requirements met by applying BAT,
but are applicable only to new production. A comparable modification in BPWTT is not possible given the definition of the technology
objective for the municipal sector. Thus, alternative policy 3 is to
meet NSPS on 1983 production by applying BPT/BPWTT rather
than applying BAT/BPWTT, which is currently mandated in the
law. These alternatives are based on 1983 production estimates

TABLE 8.5

Effects of Alternative Policy 4

	Uniform BAT/BPWTT		Alternative Policy 4			Alternative as a Percentage of Uniform BAT/BPWTT	
Region	BOD Discharge (millions of lbs/yr)	Capital Cost (millions of $)	Percentage of Counties Affected	BOD Discharge (millions of lbs/yr)	Capital Cost (millions of $)	BOD	Capital Cost
New England	67	415	40	90	211	134	51
Mid Atlantic	250	1,979	39	307	986	123	50
South Atlantic	172	1,694	1	182	1,683	106	99
Great Lakes	125	2,315	0	125	2,315	100	100
Ohio	88	2,778	0	88	2,778	100	100
Tennessee	24	230	0	24	230	100	100
Upper Mississippi	97	754	0	97	754	100	100
Lower Mississippi	58	857	16	72	464	124	54
Souris–Red–Rainy	4	9	0	4	9	100	100
Missouri	94	248	0	94	248	100	100
Arkansas–White–Red	28	1,368	0	28	1,368	100	100
Texas Gulf	45	2,659	9	61	1,948	136	73
Rio Grande	5	236	0	5	236	100	100
Upper Colorado	1	15	0	1	15	100	100
Lower Colorado	14	36	0	14	36	100	100
Great Basin	4	186	0	4	186	100	100
Pacific Northwest	81	297	2	106	231	131	78
California	165	857	15	228	517	138	60
National Totals*							
Lower–bound	1,300	17,100	1	1,400	16,700	108	98
Upper–bound	1,300	17,100	5	1,500	14,400	115	84

*Rounded.

Note: See text for definition of lower- and upper-bounds.

Source: NRDI Computer Outputs. 1975.

rather than 1973 production estimates used in the other alternatives because the justification for more stringent standards is based on the need for offsetting the effects of growth.

Limiting the application of more stringent technology to new production and allowing for a six-year increase in overall industrial production would result in industrial discharges being only slightly greater than those from 1977 production after applying BPT (see Table 8.5). Average industrial discharge reductions would be 76 and 87 percent of 1973 discharges for 1983 production after applying BPT and 1983 production after BAT, respectively. This compares with the 78 percent reduction achieved for 1977 production after BPT. Using a policy limiting more stringent technology to new production rather than all production would result in a loss of 2 percent of the abatement achieved for 1977 production after BPT. Not surprisingly, the number of basins ranked in the highest water quality category is almost the same as for 1983 production after BAT. The only exception is the loss of one basin from the highest water quality category in the Great Basin region.

Limiting more stringent reduction technology to only new residuals sources would result in a 40 percent average reduction in the BAT costs for 1983 production (see Table 8.4). The reduction in technology costs would range from a high of 65 percent for the Pacific Northwest to a low of 29 percent for the Upper Colorado. The largest absolute reduction in costs would be approximately $1 billion for the Great Lakes and the smallest absolute reduction in costs would be less than $0.1 billion for the Upper Colorado.

In summary, an alternative limiting more stringent technology to new industrial production would approximately maintain the discharge level of 1977 production after applying BPT rather than allowing discharge to increase to a level approximating 1973 conditions. This limitation would apparently not result in a noticeable change in water quality rankings, but would result in a $6 billion savings in industrial technology costs for the nation.

ALTERNATIVE POLICY 4: EXCLUDING OCEAN DISCHARGERS FROM UNIFORM BAT/BPWTT OBJECTIVES

As with inland basins where the assimilative capacity of surface receiving waters may make it possible to achieve desired water quality after applying BPT/ST, and where applying BAT/ BPWTT has little or no additional effect on water quality, in certain coastal areas the assimilative capacity of the oceans as receiving waters is also an important consideration.

The waste assimilative capacity of the oceans is great, because, "An enormous amount of water is moved along the United States shores by major coastal currents. Typical transport rates for the California current or Florida current, which range from 10 to 60 million cubic meters per second, are millions of times greater than the flow of the largest wastewater dischargers. Because of this great excess of available dilution water, the dilution of discharged wastes depends only on the rate of mixing of ocean and wastewaters" (Southern California Coastal Water Research Project, 1975). Consequently, it is worthwhile to investigate the alternative of excluding from BAT/BPWTT objectives those areas that already have or will have the potential of ocean discharge from BAT/BPWTT objectives.

It is difficult to realistically assess the potential benefits of this alternative since estuarine areas vary significantly along the coasts. In some wetland areas, potential damages from ocean outfalls may necessitate applying BAT/BPWTT objectives; while in others, coastal currents and tidal actions may dilute residuals discharges sufficiently to obviate the need for more stringent treatment.

To demonstrate this alternative, the range of possible regional cost savings is illustrated by computing a reasonable approximation. The lower bound, alternative 4A, is applied in three water resource regions, South Atlantic (I as shown in Figure 8.1) California-South Pacific (II), and Columbia-North Pacific (III), in which ocean discharge is either already practiced or known to be feasible. The upper bound, alternative 4B, is the cost savings if all coastal areas in the continental United States were to substitute ocean dilution for meeting BAT/BPWTT requirements. Coastal areas in the three water resource regions representing the lower bound are defined as those coastal counties having a treatment plant or plants determined by EPA not to require secondary treatment. Coastal areas for the upper bounds are defined to be all U.S. coastal counties, excluding those already excluded in the three regions mentioned above. For both upper and lower bounds it is assumed that all point sources in a county take advantage of substituting ocean discharge for BAT/BPWTT.

Analyzing this alternative yields some interesting, but not unexpected, results (see Table 8.6). At the national level the possible savings range from $0.4 to $2.7 billion, or from 2 to 16 percent of total uniform BAT costs. However, at the regional level the range of savings becomes greater. Of the three regions considered in the lower bound, two show significant savings; California-South Pacific, with a savings of $340 million, a 94 percent reduction from uniform BAT costs; and Columbia-North Pacific, with a savings of $66

TABLE 8.6

BOD Discharges and Costs for Alternative Policy 4

	Alternative Policy 4A				Alternative Policy 4B			
Region	BOD Discharge (millions of pounds per year)	Capital Cost (millions of dollars)	Percent of BOD after BPT/ST	Percent of BPT/ST Cost	BOD Discharge (millions of pounds per year)	Capital Cost (millions of dollars)	Percent of BOD after BPT/ST	Percent of BPT/ST Cost
New England	*	*	*	*	90	211	96	12
Mid Atlantic	*	*	*	*	307	986	70	15
South Atlantic	182	1,683	53	38	182	1,683	53	38
Great Lakes	*	*	*	*	*	*	*	*
Ohio	*	*	*	*	*	*	*	*
Tennessee	*	*	*	*	*	*	*	*
Upper Mississippi	*	*	*	*	*	*	*	*
Lower Mississippi	*	*	*	*	72	464	65	34
Souris–Red–Rainy	*	*	*	*	*	*	*	*
Missouri	*	*	*	*	*	*	*	*
Arkansas–White–Red	*	*	*	*	*	*	*	*
Texas Gulf	*	*	*	*	61	1,948	32	72
Rio Grande	*	*	*	*	*	*	*	*
Upper Colorado	*	*	*	*	*	*	*	*
Lower Colorado	*	*	*	*	*	*	*	*
Great Basin	*	*	*	*	*	*	*	*
Pacific Northwest	106	231	83	24	106	232	83	24
California	228	517	71	18	228	517	71	18
National Totals	1,398	16,683	52	42	1,508	14,382	56	36

*Alternative is not applicable--results are the same as BRT/BPWTT.

Note: Alternative 4A is lower bound case, 4B is upper bound case.

Source: NRDI Computer Outputs. 1975.

FIGURE 8.1

Identification of Coastal Counties Assumed to Substitute Ocean Dilution For BAT/BPWTT

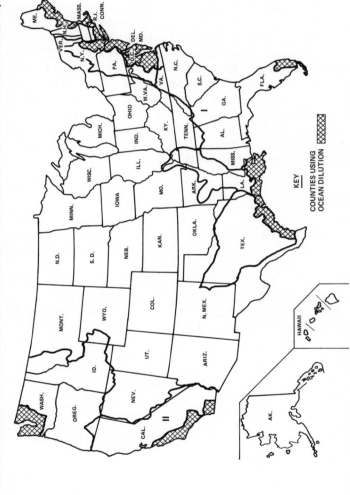

KEY

COUNTIES USING
OCEAN DILUTION

Source: Compiled by the authors.

million, a 22 percent reduction from uniform costs. In addition,
the four other regions considered in the upper bound achieve substantial savings, ranging from almost $1 billion, or 50 percent
reduction of total uniform BAT costs, in the Middle Atlantic region
to $204 million, or 40 percent reduction of total uniform BAT costs,
in the New England region.

Admittedly, the savings estimated for the four regions used
in the upper bound are exaggerated; all their coastal counties are
assumed to use ocean dilution. On the other hand, the savings
estimate for the regions used in the lower bound are believed to
be reasonable since only those counties are omitted in which EPA
has indicated that municipal treatment facilities are not required to
go to secondary treatment. Although the range of possible savings
at the national level is modest, the substitution of ocean dilution
for meeting uniform BAT objectives in certain regional areas will
produce significant regional savings, a further illustration of the
potential benefits of regional diversity.

REFERENCES

National Research Council. 1975. National Residuals Discharge
        Inventory Computer Outputs. Washington, D.C.: National
        Research Council.
Southern California Coastal Water Research Project. 1975. Environ-
        mental Effects in the Disposal of Municipal Wastewaters in
        Open Coastal Waters. Southern California Coastal Water
        Research Project.

9

SUMMARY OF FINDINGS, AND CONCLUSIONS

The development of the various segments of the NRDI has given the authors the opportunity to acquire a broad view of the water quality problems facing the nation, the solutions that have been legislatively mandated by PL 92-500, and the mechanisms through which EPA has implemented the law. From this perspective, it is clear that prior studies, including those of NCWQ, have taken a narrow view of the problem.

In Chapter 1, we argued that a full and complete investigation should have included consideration of the uniformity, finality, and enforceability provisions, the essential components of the 1972 Law. Instead, in their study program, NCWQ restricted their mandate to the finality and enforceability provisions and ignored the uniformity provision. Without information about the consequences of the uniformity provision, making a socially wise midcourse correction is impossible. The purpose of this analysis was to fill this void so that a recommendation concerning the potential midcourse corrections could be developed based on quantitative analysis of the many facets of the 1972 Law.

This chapter first presents a summary of the findings of the study, concisely reiterating much of Chapters 7 and 8, and then presents the prinicipal conclusions drawn from the analysis.

FINDINGS

National Consequences of the 1972 Law

Uniform implementation of the 1977 and 1983 technological objectives would result in a significant reduction of point-source discharge into the nation's waters. However, there would still remain significant discharge from areal sources, which have markedly different regional variations.

Point-Source Residuals

Uniform implementation of the 1977 and 1983 technological objectives would result in a significant reduction of BOD and SS discharge from point sources. The total point-source loadings of BOD would decrease by 75 and 87 percent after implementation of BPT/ST and BAT/BPWTT, respectively, based on 1973 discharge levels. Similarly, total point-source loadings of SS would decrease by 98 percent after both BPT/ST and BAT/BPWTT. In addition, municipal point-source loadings of P would decrease by one-third after ST and an additional one-sixth after BPWTT. Municipal point-source loadings of N would decrease by 10 percent after ST and remain essentially the same after BPWTT.

Technology Costs of Controlling Point Sources

Implementation of the 1977 technological objective by industrial and municipal activities is estimated to require a capital investment of approximately $45 billion. The NRDI has the ability to directly estimate $11.7 billion in industrial capital costs and $28.4 billion in municipal capital costs. The other $4.8 billion in industrial costs is an approximated capital cost for pretreatment of incompatible wastes (costs that were not included because of the uncertainties associated with the pretreatment requirements and the fact that the wastes are discharged into municipal systems) and treatment of other discharges not covered in the NRDI.

The $16.5 billion in industrial costs is less than half the $44 billion estimated by the NCWQ, but is approximately equal to the $21 billion estimated by the Council on Environmental Quality (CEQ) and the $16 billion estimated by EPA. The $28 billion estimate of municipal costs is approximately equivalent to the NCWQ estimate of $36 billion, which was based on serving the 1973 population and included interceptor sewer costs. Both estimates are lower than the EPA report to Congress on the Needs Survey, which reported combined treatment and interceptor costs of $53 billion, for the 1990 population.

Implementation of the 1983 technological objective by industrial and municipal activities is estimated to require a capital investment. The NRDI has the capacity to estimate

$6.8 billion in industrial capital costs and $10.3 billion in municipal costs. The other $2.9 billion is an approximated capital cost for pretreatment of incompatible residuals in industry.

The $10 billion in industrial costs is less than one-third of the $30 billion-$36 billion estimated by the NCWQ and is slightly less than the $14 billion estimated by CEQ. EPA has not officially estimated the costs of the 1983 technological objective.

Growth Effects

Changing the production and population basis for estimating residuals and costs does not significantly after most of these estimates. The BPT-level residual estimate based on 1973 production is only 4 percent less than it would be if based on 1977 production, and the BAT-level residual estimate based on 1973 production is 25 percent less than it would be if based on 1983 production. In either case, the 1983 discharge would only be approximately 10 percent of the 1973 discharge. The ST-level residual estimate based on 1973 population is only 12 percent less than it would be based on 1977 population and the BAT-level residual estimate based on 1973 production is 36 percent less than it would be if based on 1983 population. In either case, the 1983 discharge would only be approximately 20 percent of the 1973 discharge. Similarly, the BPT cost estimate based on 1973 production is only 11 percent less than it would be if based on 1977 production, and the BAT cost estimate based on 1973 production is 50 percent less than it would be if based on 1983 production. While a 50 percent difference is significant, this difference is exaggerated by assuming that all new growth occurs in new plants. The ST cost estimate based on 1973 population is 7 percent less than it would be if based on 1977 population, and the BPWTT cost estimate is 28 percent less than it would be if based on 1983 population.

Uncontrolled Sources

The application of BPT/ST and BAT/BPWTT would signifi-cantly reduce BOD and SS residuals from controlled sources, but would leave unaffected other significant sources. The control of point sources would reduce their 1973 discharge of BOD and SS by 87 and 98 percent, respectively, and the total 1973 discharge, including areal sources, by 44 and 36 percent, respectively. Areal source-BOD would then be approximately eight times larger than point-source BOD, and areal-source SS, 100 times larger than point-source SS.

Similarly, the application of ST and BPWTT would significantly reduce P and N residuals from municipal activities, but
would not affect other significant sources of the same residuals.
The control of municipal activities would reduce the total 1973
discharge, including that of areal sources, by 15 and 5 percent,
respectively. Nonirrigated agricultural activities would discharge
three times more P than municipal activities and twice as much
N.

While the 1972 Law's emphasis on controlling point source
does leave significant uncontrolled residuals from areal sources,
residuals from these sources might not have the same impact on
water quality as residuals from point sources. As explained earlier,
residual discharges of BOD and SS from urban activities occur
during storm events, which also increase the stream flow and, thus,
residual assimilative capability. This combination usually results in
only a temporary impairment of water quality. Similarly, residual
discharges of BOD and SS from nonirrigated agricultural activities
occur during storm events and are diffuse rather than concentrated
in nature, thus reducing their impact. However, nutrient residuals
from areal sources could be as significant as those from point
sources if they accumulate in receiving water and remain available
to stimulate algal growth. Uncontrolled activities would be a significant source of residuals, but probably not as significant as indicated
by the magnitude of the discharge.

Regional Consequences of the 1972 Law

Aggregation of NRDI data by regions reveals the major geographic differences in the relative magnitude of residuals discharges
and technology costs for major activities, which are masked by
national totals. Similarly, it reveals that the magnitude and type of
uncontrolled residuals varies considerably by region in the United
States. Lastly, data on changes in relative water quality can be
meaningfully presented at this level of aggregation.

The magnitude of residuals discharges and the replacement
value of the in-place residual reduction technology varied considerably in 1973 and will continue to vary in the future. In seven
of 18 regions industrial activities contributed a majority of the total
point-source BOD discharge. Two East Coast regions (Middle and
South Atlantic), one Midwest region (Great Lakes), and one Far West
region (California) accounted for approximately 80 percent of the
total point-source BOD discharges. In all regions industrial activities contributed a majority of the total point-source SS discharges.

In all regions the replacement value of municipal residual reduction technology exceeds the replacement value of industrial residual reduction technology, indicating that there has been a greater investment in municipal waste treatment systems than in industrial. The replacement value of industrial technology in six regions accounts for 75 percent of the total industrial investment.

The projected net residuals and costs of BPT/ST continue to show considerable regional variation. In all regions the municipal costs of achieving ST would be greater than the industrial costs of meeting BPT. After this capital investment, industrial activities would account for a majority of the total point-source BOD discharge in only five regions. However, this investment would not change industrial dominance of SS discharge in all regions.

After installation of BAT/BPWTT, industrial costs would be greater than municipal costs in eight regions. The notable exceptions, where the municipal investment is twice as great as the industrial investment, are the Arkansas-White-Red, Texas Gulf, and Rio Grande regions. Four regions account for 60 percent of the projected BAT/BPWTT costs. After this investment, industrial activities would account for a majority of the total point-source BOD and SS discharge in 4 and 18 regions, respectively.

In 1973 BOD from point sources was greater than from urban runoff sources in all regions and SS from point sources was greater than from urban runoff sources in 12 regions. Similarly, BOD from point sources was greater than from nonirrigated agricultural sources in 13 regions and SS from point sources was greater than from nonirrigated agricultural sources in five regions.

After the application of BPT/ST, point sources of BOD and SS would no longer be the dominant source in many regions. BOD from urban runoff sources would be greater than from point sources in eight regions, primarily in the East and Midwest. Similarly, SS from urban runoff sources would be greater than from point sources in 12 regions, excluding only those regions in the West. BOD from nonirrigated agricultural sources would be greater than from point sources in 13 regions, excluding only three eastern regions, and the Great Lakes, and Texas Gulf regions. SS from nonirrigated agricultural sources would be greater than from point sources in all 18 regions.

After the installation of BAT/BPWTT, this trend continues. BOD from urban runoff sources would be greater than from point sources in 12 rather than eight regions, again primarily in the East and Midwest. However, SS from urban runoff sources would not increase in dominance over that from point sources after BPT/ST. SS from nonirrigated agricultural sources would be greater than

from point sources in 14 rather than 13 regions, with the shift
occurring in the Great Lakes region. SS from nonirrigated agricultural sources would continue to dominate that from point sources
in all 18 regions.

Municipal sources accounted for a majority of the total P
discharge in 10 regions, primarily in the East and Far West. Similarly, municipal sources accounted for a majority of the total N
discharge in four regions, the three eastern ones and the Great
Lakes.

Relative Water Quality Impacts

Under 1973 residual discharge conditions, the residuals
dilution index suggests that water quality during low flow conditions
varies considerably among and within regions. Only in the Lower
Mississippi, Upper Colorado, and Pacific Northwest regions were
all basins in the least degraded class. Among all 99 basins, 52
were in the least degraded category. Most of the regions had at
least 50 percent of their basins in the two classes indicating the
least degradation, but there were also several basins within those
regions in which the water quality might be severely degraded by
the 1973 level of discharge.

Regulation of point sources to achieve BPT/ST would have a
noticeable impact upon water quality in most regions as measured
by the water quality index. In 13 regions, over 70 percent of the
basins within the region would achieve water quality within the
least degraded class, and 80 basins among all the regions would
fall into the same class. In 16 regions, over 70 percent of the
basins within the region, and 87 basins among all the regions,
would achieve water quality within the three classes indicating the
least degradation. However, there would still remain several basins
in which the water quality index remains very high and these basins
are scattered throughout the country.

Regulation of point sources to achieve BAT/BPWTT would
have only a limited additional impact upon water quality in most
regions as measured by the index. In 13 regions, as is the case
after BPT/ST, over 70 percent of the basins within each region
would achieve water quality within the least degraded class, but
87 basins rather than 80 basins would fall into the same class. In
four eastern and midwestern regions one basin would shift from the
second least to the least degraded class. In 16 regions, as is true
after BPT/ST, over 70 percent of the basins within each region
would achieve water quality within the three classes indicating the

least degradation and 93 basins among all the regions would fall into those classes. A noticeable change is the reduction of the number of basins in the classes indicating severe degradation of water quality. Examples of this type of change are found in the Great Lakes, Arkansas-White-Red, and Great Basin regions. However, there would remain basins in several regions that would still have severe water quality degradation even with the achievement of BAT/BPWTT.

Since the preceding discussion is based upon the low flow conditions as defined in Chapter 7, it therefore overstates the relative water quality attainable from controlling point sources. Under average flow conditions, when both point and nonpoint discharges are considered, the relative improvement in water quality (river basins achieving higher rankings) after BPT/ST and BAT/BPWTT would not be as great. After BPT/ST, 64 rather than 80 basins would be in the least degraded class and 80 rather than 90 basins in the three least degraded classes. Similarly, after BAT/BPWTT, 69 rather than 87 basins would be in the least degraded class and 81 rather than 93 basins in the three least degraded classes.

Alternatives to the Uniform Application of BAT/BPWTT

The principal findings of the analyses are that the uniform application of BPT/ST objectives, through wasteful of national capital resources in a few situations, would achieve considerable results, and that the uniform application of BAT/BPWTT objectives would result in little change in either residual discharges or relative water quality despite a large additional capital investment. In light of these findings, four alternatives to the uniform application of more stringent, BAT/BPWTT objectives have been examined.

Alternative policy 1 requires that the BAT/BPWTT objectives be met only in those basins that have relatively severe water quality problems. The application of BAT/BPWTT is limited to basins that have a BOD dilution index greater than or equal to 3.0 milligrams per liter. The results show that if uniform water quality is a policy objective, it can be obtained at a 72 percent reduction of the costs of uniform BAT/BPWTT if the more stringent effluent limitations are applied to only 21 instead of all 99 basins. The results suggest that in the remaining 78 basins, BAT/BPWTT may not be necessary because they generally may have met water quality standards after BPT/ST. Basins with no additional cost would be located in the New

England, Tennessee, Upper Mississippi, Lower Mississippi, Upper
Colorado, and Pacific Northwest regions.

The second alternative policy is to require only those residuals
sources that have the least cost to invest in more stringent technology
in order to achieve the residual reduction accomplished by uniform
BAT/BPWTT. It is based on considering all major sources of residuals, including urban runoff and nonirrigated agriculture, as well as
applying both BPT/ST and BAT/BPWTT nonuniformly in order to
achieve a given level of residual reduction at the least cost.

The results showed that a 33 percent reduction in total BPT/ST
and BAT/BPWTT costs, 60 percent in BPT/ST costs and 10 percent
in BAT/BPWTT costs, can be obtained, and the same quantities of
residuals removed, by substituting a nonuniform cost-effective policy
approach for uniformity. The five regions that benefit the most from
a cost-effective approach are the Upper Mississippi, Lower Mississippi, Missouri, Rio Grande, and Arkansas-White-Red.

This policy suggests that if the quantities of residuals to be
removed are the same as those removed after applying BPT/ST
and BAT/BPWTT, then the objectives can be achieved at lesser
costs by pursuing a cost-effective approach.

The third alternative policy is to require that only new industrial production meet BAT/BPWTT objectives, thereby requiring
1973 production levels to meet only BPT objectives. This alternative is based on future production estimates rather than 1973 production because the justification for more stringent standards is
based on the need for offsetting the effects of growth. Limiting more
stringent technology to new production and allowing for a six-year
(1977-83) increase in overall industrial production would result in
industrial discharge being approximately equal to that from
1977 production after BPT/ST for the nation. More importantly,
limiting more stringent technology to only new sources would on
average result in a 40 percent reduction in the 1983 BAT/BPWTT
costs. The reductions in technology costs would range from a high
of 65 percent for the Pacific Northwest region to a low of 29 percent
for the Upper Colorado region. This alternative illustrates that if
the quantities of residuals removed after uniform BPT/ST are sufficient to meet water quality standards, requiring more stringent
controls only on new production would maintain 1977 gains and save
additional industrial capital investment.

The fourth alternative policy is not to require point-source
dischargers in all counties to meet the BAT/BPWTT objectives if
they have the potential for ocean discharge. This alternative is
based on computing a lower bound, which excluded point-source
dischargers in selected counties in only three regions with generally recognized waste assimilation capacity, and an upper bound,

which excluded point-source dischargers in selected counties in regions with potential tidal dilution capacity. The lower bound resulted in a national cost saving of $0.4 billion, or only 2 percent of the NRDI estimated BAT/BPWTT costs, and the upper bound resulted in a national cost savings of $2.7 billion. This alternative illustrates that if the nation is willing to use the natural assimilative capacity of the oceans, a few regions would achieve a significant reduction in BAT/BPWTT costs.

In summary, the four alternatives to the uniform application of the BAT/BPWTT objective would result in saving between 2 percent and 70 percent of the costs of meeting uniform BAT/BPWTT without a significant deterioration of either the residual reductions or the water quality gains achieved with the BPT/ST objective. Since these alternatives are not mutually exclusive, all or part of them could be adopted simultaneously and result in a significant percentage reduction in the costs of uniform BAT/BPWTT.

CONCLUSIONS

The preceding summarizes many of the major consequences of pursuing a water quality management policy that first applies BDT/ST and then BAT/BPWTT uniformly to all point sources, yet excludes other residuals sources. On the basis of these results we have arrived at the following conclusions about the BAT/BPWTT objective.

The first conclusion is that uniform application of increasingly stringent point-source controls is an inefficient method of allocating resources to solve the nation's water quality problems.

Some of the proposed alternative policies suggest that taking advantage of certain regional characteristics, such as waste assimilation capacity, could produce a 70 percent savings in estimated BAT/BPWTT costs and achieve comparable residuals removal results. If nonirrigated agricultural and urban runoff sources are considered and a cost-effective approach followed by first controlling activities that have the lowest cost-per-pound-of-removal ratio, further savings can be made over uniform BPT/ST and BAT/BPWTT costs. The implications of the corollary to the foregoing discussion are also important: If substantial savings can be obtained and a similar level of water quality improvement obtained through policies other than uniformity, what further improvements in water quality are possible with the savings obtained?

The second conclusion is that uniform application of increasingly stringent point-source controls would result in only marginal improvements in water quality in many basins, would be necessary to meet water quality standards in some basins, and yet be inadequate to solve water quality problems in a few basins.

While regulation of point sources to achieve BPT/ST would have a noticeable impact upon water quality, the more stringent controls associated with BAT/BPWTT would have only a limited impact. Application of BPT/ST would increase the number of river basins in the least degraded water quality class from approximately 50 to 80 percent and put 90 percent of all basins in the three least degraded classes. However, application of BAT/BPWTT would only increase the number of river basins in the least degraded class from approximately 80 to 85 percent and put 92 percent of all basins in the three least degraded classes. Installation of more stringent controls would result in some marginal gains, but would not solve water quality problems in all basins.

While the water quality index is only a rough ranking of river basin pollution problems and is based on only one parameter, the relative ranking is consistent with the general trend documented by NCWQ. In addition, the water quality trend data are reinforced by the residual trend data. Application of BPT/ST would reduce BOD loadings by 75 percent of their 1973 level and SS loadings by 98 percent. However, application of BAT/BPWTT would reduce BOD loadings by 87 percent (a 12 percent further decrease) of their 1973 level and SS loadings by 98 percent (a 0 percent further decrease).

The third conclusion is that uniform application of increasingly stringent point-source controls is not needed within the 1985 time frame of the law to offset the effects of industrial and population growth.

Reductions in residuals achieved by applying BPT/ST would not be lost if the nation failed to invest in BAT/BPWTT. Application of BPT/ST to 1977 production and population levels would reduce BOD discharges from point sources to 30 percent of their 1973 level. Without the application of BAT/BPWTT (but with new source performance standards), BOD loadings from point sources would increase by 15 percent from 1977 to 1983, but would still be only 35 percent of their 1973 level.

The fourth conclusion is that uniform application of increasingly stringent controls to only point sources would ignore other major sources of residuals and thus would probably fail to eliminate water quality problems in many river basins.

In 1973, the amount of BOD and SS residuals from point and areal sources were roughly comparable. Application of BPT/ST would result in BOD residuals from areal sources being three times greater than those from point sources. Similarly, SS residuals from areal sources would be two times as great as from point sources. Application of BAT/BPWTT would itensify the difference. BOD residuals from areal sources would be eight times as great as from point sources and SS residuals from areal sources would be 100 times greater than from point sources.

Under average flow conditions when both point and areal sources are considered, the relative improvements in water quality after BPT/ST and BAT/BPWTT are overstated in conclusion two above. Only 65 rather than 80 percent of the basins would be in the least degraded class after BPT/ST, and only 70 rather than 85 percent would be in the least degraded class after BAT/BPWTT. While similar residuals from areal sources may not have the same impact on water quality as those from point sources, their magnitude is sufficient to indicate a potentially serious problem.

The fifth conclusion is that uniform application of increasingly stringent point-source controls would result in significant regional variations in technology costs and would not eliminate regional variations in residuals discharges per capita.

In essence, implementation of nationally uniform technology objectives actually results in regional nonuniformity. Costs of upgrading from BPT/ST to BPT/ST to BAT/BPWTT show considerable variation among the 18 regions. Per capita costs of upgrading to BAT/BPWTT range from $50 to $270 with a national average of $80. While most regions experience a higher investment required by industry, municipal investment twice as great as industrial investment is required in three regions. Thus, the costs of implementing the BAT/BPWTT objectives are not close to uniform even when looking at relatively large areas such as the 18 regions.

Nor does the implementation of BAT/BPWTT eliminate the variation in per capita BOD and SS discharge. The per capita BOD discharge would be 6 pounds on the average for the nation, but would vary from 0 to 11 pounds among the 18 regions. The per capita SS discharge would be 63 on the average for the nation, but would vary from 0 to 810 pounds among the 18 regions. Including effects of uncontrolled sources such as nonirrigated agriculture and urban storm would increase the ranges considerably.

In summary, we must conclude that the administrative benefits of having uniform technology standards, which are primarily ease in developing standards and simplified enforcement, are purchased at a very high price. The national application of more stringent uniform standards to point sources increases considerably the technology costs of achieving water quality. In many areas, more stringent control is not necessary to eliminate environmental degradation or to offset the effects of industrial and population growth. This policy also detracts attention from potential problems with other residual sources. Thus, a midcourse correction in technology objectives appears necessary in order that the nation may make the needed capital investments beyond BPT/ST in a way which insures that benefits are commensurate with the costs.

10

MIDCOURSE CORRECTIONS
IN TECHNOLOGY OBJECTIVES

NCWQ PROPOSED MODIFICATIONS

The NCWQ commissioners responded to the findings in its
Staff Report to the National Commission on Water Quality* (NCWQ
1976) and public hearings by proposing modifications not only in
the 1983 requirements of the 1972 Act, but also in the 1977 requirements (Appendix A). Discussion of both sets of modifications is
necessary to understand the proposed midcourse corrections in
technology objectives.

The commissioners' proposed modifications in the 1977
requirement are discussed below.

First, they would grant extensions of time to municipal,
industrial, and agricultural dischargers in meeting the 1977 requirements, on a case-by-case basis. The NCWQ findings showed that
some point-source dischargers, primarily municipalities, would
not meet the July 1, 1977 deadline. Unless some extensions of time
were granted, these dischargers would be in technical violation of
the law. Since there are many legitimate reasons for not meeting
the deadline, this modification was necessary in order to maintain
the credibility of the water program.

Second, the NCWQ would waive, defer, or modify the 1977
requirement on a case-by-case basis where the discharger demon-
strates that adverse environmental impacts of such modification

---

*This report drew upon the many contractor reports and
other information that were available. The nearly identical draft
released for comment and review in November, 1975 is referred
to as the "Staff Report" or "Staff Draft Report".

will be minimal or nonexistent, or that the costs of the requirement are disproportionate to projected environmental gains.

This modification has the potential to lead to erosion of the 1977 requirement, a requirement which would achieve considerable environmental gains at a reasonable cost according to both the NCWQ and NRDI analyses. Such a waiver is unnecessary in a situation where the vast majority of the actions initiated as a result of the 1977 requirement would have benefits commensurate with costs. That the commissioners should make such a modification is surprising in light of the Staff Report, which continually stressed the difficulties and uncertainties of assessing environmental damages and estimating capital and operating costs. If the $17 million spent on NCWQ studies could not answer these questions, how could NCWQ expect an individual discharger to address them?

Third, the NCWQ would waive, defer, or modify the 1977 requirement, on a category-by-category basis, for near-ocean discharges of municipalities, for pretreatment by industries discharging into municipal systems, for existing municipal oxidation ponds and lagoons, and for situations where the adverse environmental impacts of such modifications will be minimal or nonexistent or where the costs of the requirement are disproportionate to the projected environmental gains (de minimus situations).

Environmental and cost findings in the Staff Report do offer some support for modifications in the ST requirement for ocean dischargers and for small communities using oxidation ponds and lagoons; administrative considerations support modification of pretreatment requirements and application of a de minimus rule. While the specificity of this set of modifications would prevent a major weakening of the 1977 requirement, at the least, the benefits of modifying ST requirements could be achieved by assigning deferred or low priority to these investments or by redefining ST requirements. The findings in the Staff Report do not document, nor does any other source, how the administrative benefits of modifying pretreatment requirements and allowing a modification rule in situations with minimal environmental impact would be greater than the potential environmental degradation if EPA did not regulate these activities. Thus, this set of modifications is largely unwarranted on the grounds either that the same objectives could be accomplished within the law or that there is insufficient evidence to support a change.

The commissioners' proposed modification of the 1983 requirement would postpone the deadline by which municipal, agricultural, and industrial discharges shall be required to meet the BAT/BPWTT requirements from July 1, 1983 to a date not less than five and no more than ten years after 1983. In the interim period, NCWQ suggested several other actions including upgrading of BPT effluent limitations and more stringent controlling of residuals from urban runoff and agriculture, where cost effective.

NCWQ apparently based this modification on the following reasons. First, municipalities would barely comply with the ST requirement within that period, let alone the BPWTT requirement, and industries would have a difficult time complying particularly with BAT. Second, industry would bear a substantial cost in meeting BAT requirements even though that cost would be less than the BPT cost. Third, a delay in point-source discharge requirements would allow consideration of cost-effective solutions, that is, control of urban runoff and agriculture. Fourth, and seemingly most important, data were not available to determine whether environmental benefits from applying BAT/BPWTT would be commensurate with the costs. Until BPT/ST has been widely installed, data are too limited to accurately assess the improvements in water quality resulting from even more stringent requirements.

These reasons do not make a strong case for a lengthy delay in the 1983 requirements, a delay which may make the 1983 requirement meaningless. Postponing BAT/BPWTT for five to ten years, probably the latter, would free point-source dischargers from the very real threat of having to make substantial capital investments in pollution control technology. Without this negative incentive, industrial dischargers would no longer have a compelling reasons to participate in water quality planning emphasized in the law or to invest systematically in in-plant modifications to increase productivity while reducing residuals discharges. Similarly, municipal dischargers would no longer have a compelling reason to participate in water quality planning or to adopt waste load modification measures.

The question is, would sacrificing the negative incentive that sustains the interests of point-source dischargers in water quality management be worth the above benefits enumerated by NCWQ? On the whole, the answer is no.

First, delaying the BAT/BPWTT deadline is not necessary in order to prevent point-source dischargers from violating the law. Evidence from NCWQ contractors (Energy and Environmental Analysis 1975) and elsewhere suggest that industry will generally meet the 1977 deadline. Even allowing for some slippage, industry would have three to four years to meet the 1983 deadline. While many large municipalities would not meet the 1977 deadline, they could meet the 1983 deadline because EPA has defined BPWTT to be equivalent to ST. Since the vast majority of municipalities would meet the ST objective in the mid-1980s, they would automatically

meet the BPWTT objective at the same time. Thus, there is no
need to postpone the 1983 deadline on account of the inability of
industries and municipalities to install the appropriate technology.

Second, both the benefits of avoiding substantial industrial
costs for BAT and of considering potential cost-effective measures
can be achieved by ways other than delaying the 1983 requirement.
These benefits could be achieved by exercising the water quality
planning provisions of the law and by allowing the findings of local
and regional planning to determine the type and magnitude of future
investments in waste treatment technology. The actual dimensions
of such a program are discussed in more detail later in this chapter.

Third, there are few reasons to think that sufficient data on
environmental benefits, and on actual industry expenditures rather
than government-estimated costs of waste treatment technology,
would be available by the mid-1980s. The problem is particularly
critical in assessing environmental benefits, which requires understanding the relationships among waste discharges, ambient water
quality, and resulting benefits. Even if society gains adequate knowledge about the fate of wastes such as toxic materials in the environment, which is unlikely, there would remain the problem of assessing
the dollar benefits of water quality improvements. In the end,
determining whether the costs of water pollution control are commensurate with the benefits will depend on political judgement,
whether at the local, state, or national level.

The possibility of having even minimal data by 1980 would
probably be eliminated if industrial and municipal dischargers were
no longer faced with the negative incentive of the 1983 requirement.
Without industry's cooperation and investment in research in this
area, and without the pressure exerted by industry and local and
state governments on the federal government to invest in research
in this area, we cannot envision that the nation will be in a sufficiently better position in the 1980s to determine scientifically the
benefits and costs of water pollution control. With the activity in-
duced by the negative incentive, there is the possibility that a
better (albeit not sufficient) data base would be available to influence
the political decision that determines whether water pollution control
is a wise use of national resources.

EPA RESPONSE

The EPA response to NCWQ's proposed midcourse corrections
is to defend the uniform application of a more demanding level of
technology by the 1983 deadline (Quarles 1976 and Alm 1976). According to the EPA, the NCWQ underestimated the benefits of the

BAT/BPWTT objective and overestimated the time delays in meeting the 1977 and 1983 requirements. EPA contends that the application of BAT/BPWTT is still a wise investment of resources and feasible within the 1983 time frame. In addition, they claim it is the only administrative way to achieve water quality.

EPA rejects the need for midcourse corrections, first, on the basis of NCWQ benefit assessments. EPA claims that the benefits of BAT have been underestimated in the Staff Report. "The basic failure of the Staff Draft Report in estimating the benefits of a second level of technology-based standards is its concentration upon dissolved oxygen and fecal coliform as the primary indices of water quality improvement." EPA states that the BAT will require removal of other-than-easily-degradable organic material and sediment, such as nutrients, heavy metals, and organic wastes, which are of concern to human (and other species') health. Since the dangers of these wastes do not lie in depletion of oxygen, their removal is not addressed by the BOD controls achieved by BPT and BAT. Moreover, even considering only the "traditional parameters" that the Staff Report addresses, the additional removal achieved by BAT/BPWTT is substantial in light of the increased population and industrial growth over the period.

While the EPA critique of the NCWQ benefit assessment has some merit, it is not an adequate defense of the necessity for retaining the uniform application of the BAT/BPWTT requirement. One limitation of the EPA argument is the claim that the second level of technology is intended to control primarily waste parameters other than the traditional ones. In some cases, such as meat packing, organic chemicals, and petroleum, the technologies do primarily control other parameters. However, in other cases, such as iron and steel, and pulp and paper, the technologies' primary effect is additional removal of the sediment and/or organic material. Thus, comparing the benefits from controlling traditional parameters and costs is a legitimate exercise in many cases. The EPA goes on to note, in response to the NCWQ's findings, that there is "substantial" removal of traditional parameters with the use of BAT, and supports the position that BAT/BPWTT is necessary to offset the effects of industrial and population growth. EPA's support of this position is somewhat contradictory to their first point, which is that the primary purpose of BAT is not to remove traditional parameters. Further, the argument is incorrect according to the NRDI analysis, which showed that BAT/BPWTT is not needed to offset the effects of industrial and population growth. Lastly, the EPA position is an inadequate defense of the uniform application of technology because nowhere does EPA, either in their comments on the NCWQ findings or in

other documents, offer cost and benefit data, which would support
either uniformity or their position in general. While EPA suggests
that there are additional benefits, it does not offer any empirical
evidence about the magnitude of those benefits. Nor does it present
any data on how the costs of controlling other-than-traditional
parameters (including possible additional costs if BAT requirements
are revised) affect the total costs of BAT/BPWTT as estimated by
NCWQ.

EPA rejects the necessity of a midcourse correction, second,
on the basis of delays in implementing the law. EPA claims that
"the Staff Draft Report has overestimated the delays experienced
in implementing the first phase of control and has assigned undue
significance to the delays which have been encountered with respect
to the 1983 deadline." As regards the 1977 deadline, EPA has
issued permits for industrial and municipal dischargers in the
absence of formal effluent limitations. EPA contends that the low
rate of expenditure of industry to date on pollution control is an
effort to avoid unnecessary interest and operating costs rather than
an indication of their inability to meet the 1977 requirement. As
regards the 1983 deadline, the second round of permits to industrial
and municipal dischargers is scheduled to be issued in December
1979 at the latest. Point-source dischargers would have three and
a half years beyond that time to comply with BAT/BPWTT. "Thus,
even deferring BPT/ST until 1979/1980 does not mean that 2 or 3
years must be tacked onto 1983."

While we did not conduct an independent investigation of the
time required for industries and municipalities to install BPT/ST
and BAT/BPWTT, we would agree with the EPA position. As mentioned earlier, the findings of one NCWQ contractor (Energy and
Environmental Analysis 1975) suggested that the vast majority of
industrial dischargers would meet the 1977 deadline. The low rate
of expenditure to date on the part of industry, combined with their
acknowledgment that most of them will meet the 1977 deadline, can
easily be interpreted as indicating that NCWQ has seriously overestimated the cost of meeting BPT. EPA has supported elsewhere
the desirability of delaying the ST deadline especially in light of the
nonavailability of federal funds. Since EPA has defined ST to be
equivalent to BPWTT, there is no need to delay the 1983 municipal
deadline because the vast majority of large municipal dischargers
will meet the ST requirements in the mid-1980s.

EPA rejects the need for a midcourse correction, third, in
light of the feasibility of alternative control methods. The alternative control method considered by EPA is the use of water quality
standards to determine the amount of residuals discharged into a

watercourse. EPA states that the use of water quality standards is incomplete and imperfect. They are incomplete because many state standards do not address the pollutants that are of concern once the basic traditional parameters are controlled; they are imperfect because of the methodological difficulties of translating ambient water quality levels into precise and defensible effluent limitations in individual permits.

While we share some of EPA's concern about the limitations of relying on water quality standards, we do not find these concerns a sufficient argument for the necessity of uniformly applying the second phase of technology-based effluent limitations. One reason that it is an inadequate defense is that approximately the same amount of residuals reduction accomplished by uniform application can be achieved by more selective application of controls. The NRDI analysis of alternative policies showed that limiting more stringent technology to new industrial growth rather than applying it to all industrial activity essentially maintained the residuals reduction achieved by BPT. Considering selective control of point and areal sources could achieve the same residuals reductions at a considerable cost savings. The other reason is that the methodology and data are available for relating ambient water quality to effluent limitations for many river basins, with some confidence in the results. When studies by GAO, EPA, the Corps of Engineers, and NCWQ all show essentially the same results for the Merrimack River, these results cannot be rejected as "imperfect." Nor when the NCWQ field investigations showed no perceptible improvement in water quality in segments of the Mississippi River as a result of either BPT/ST or BAT/BPWTT, can these results be rejected as "imperfect" indicators of the necessity for more stringent control of residuals dischargers. Thus, there would appear to be alternatives to the uniform application of technology on the basis of fixing residuals reduction targets and selective use of ambient water quality modeling.

In summary, the NCWQ proposed midcourse corrections would significantly weaken the BPT/ST requirements and would potentially make the BAT/BPWTT requirement meaningless. Selectively delaying the 1977 deadline is unnecessary. However, case-by-case exemptions on environmental grounds might start a significant erosion in the potentially significant gains from applying BPT/ST, and the desirable results from category-by-category exemptions could be achieved within the law. Delaying the date for implementing BAT/BPWTT for as much as ten years would make it meaningless as a negative incentive for point-source dischargers. Without this incentive, point-sources dischargers would have no significant pressure insuring their continuing interest in achieving water quality. Moreover, the gains from delaying implementation as enumerated by the NCWQ can be achieved with only slight modifications of the law, or are illusory. The benefits of saving substantial industrial costs and consideration of cost-effective measures can be achieved by relying on the water quality provisions of the law. The possibility that there will be, by the 1980s, sufficient data about environmental benefits to determine scientifically that benefits are commensurate with costs is so remote that it is not worth considering. The final balancing of benefits and costs will always be a political decision.

While the NCWQ midcourse corrections appear to have gone beyond the findings in the Staff Report, the EPA response to the NCWQ modifications appears to have ignored NCWQ's findings and potential alternatives to uniform technology applications. Although the Staff Report findings clearly did not identify some of the significant benefits associated with BAT/BPWTT, they clearly indicated that in some cases the gains from uniform application of more stringent technology are marginal. EPA ignored these findings by concentrating solely on those cases where benefits were under-estimated. The EPA response also stated there is no feasible alternative to the second phase, that of technology-based standards. In reality, approximately the same results can be achieved by selective application of technology and, in some basins, effluent limitation based on water quality is a feasible alternative to effluent limitation based on available technology.

RECOMMENDATION: REAFFIRM PLANNING PROVISIONS OF LAW

The one major recommendation based on this study is not a new step; rather it is a step that is potentially in the 1972 Law. The recommendation is simply that the Congress reaffirm the planning provisions of the law to the extent that they become the instruments for directing the midcourse correction. The BAT/BPWTT objective would be retained, but areawide planning authorities would be given the opportunity to demonstrate that more stringent uniform point-source control is not needed to maintain water quality or that there are more cost effective courses of control.

The 1972 Law permits three types of comprehensive planning for geographic areas. One is river basin planning for water quality management under Section 303(e). Basin plans are being used by the federal and state governments to determine the need for more stringent effluent limitations for point-source dischargers and the priority of grants to municipal dischargers for construction of sewage treatment plants. Another type of planning is areawide waste treatment management plans under Section 208. These plans are to become the basis for implementing local strategies for the prevention and abatement of water pollution. There is a third type of comprehensive plan under Section 209, which provides for the development of river basin plans for large geographic areas such as the Great Lakes.

The first two types of planning are now being combined by the EPA to insure that a water quality plan and implementation program are completed for all areas in each state (Federal Register 1975). The proposed regulations establish two phases of plans. Phase 1 plans continue the emphasis on point-source control and must be completed by July 1976. Phase 2 plans combine basin and areawide planning concepts into a plan that continues to update the point-source requirements and examines the effects of areal (non-point) sources.

The following elements would probably emerge from phase 2 areawide plans and implementation programs, if successful:

1. Identification of all major residual discharging activities and quantities of residuals discharged, including nutrients and heavy metals; this phase would include consideration of urban runoff, irrigated and nonirrigated agriculture, and other types of areal sources;

2. Determination of the amount of residuals discharge reduction needed to meet water quality standards;

3. Development of a cost-effective strategy for eliminating or preventing violations of water quality standards, whether violations result from point or areal sources;

4. Formulation of an implementation program for carrying out a cost-effective strategy.

On the basis of these plans and programs, three courses of action would be open to all geographic areas in the United States. One course of action would be that no additional capital investment in abatement technology for any source would be necessary because there would be no violation of water quality standards. The areawide plan would have to demonstrate to the satisfaction of the general public and local, state, and federal governments that the uniform application of BAT/BPWTT was not necessary to maintain water quality. The second course would be a cost-effective plan that identified particular residuals sources and the level of control needed to meet water quality standards. These sources could be either point or areal activities, and the degree of control could

be BAT/BPWTT for only selected activities or new source performance standards for only new industrial growth. Again the
areawide plan would have to demonstrate to the satisfaction of the
general public and local, state, and federal governments that the
uniform application of BAT/BPWTT was not necessary to maintain
water quality. The third course of action would be where the areawide plan failed to demonstrate that BAT/BPWTT were not essential
to meeting water quality standards. In that case, point sources would
have to apply uniformly BAT/BPWTT, and control would be required
for areal sources, where needed.

There are two compelling reasons for adopting the recommendation that areawide plans become the basis for midcourse corrections
that depart from the uniform 1983 requirements. First, analysis in
this study supports the cost effectiveness of exemptions from uniform
requirements. The diversity of residuals management problems and
the potential cost savings from pursuing other-than-uniform policies
would appear to be sufficient justification for establishing some
institutional arrangement for securing exemptions. Second, if water
quality management is constrained to controlling only point sources
more stringently, the nation may well be forced to invest substantial
amounts of capital in technology that fails to deal with significant
sources of residuals or that fails to secure appreciable residuals
reductions or water quality improvements.

The justification for retaining the BAT/BPWTT objective is
based on the perception that neither municipal nor industrial dischargers would take the planning process or implementation programs seriously unless motivated by a legal requirement for
additional capital investment. There is little evidence in the history
of water quality planning under the 1965 Law (Section 8) that those
sources which should have been controlled were actually motivated
to action as a result of planning (General Accounting Office 1969).
The same course of nonaction may well occur as a result of the new
round of plans in light of the general weakness of legal sanctions and
a total absence of investment requirements to participate seriously
in areawide planning.

Even if the final EPA areawide planning regulations contained
the proposed legal sanctions to deal with inadequate plans or the
failure to implement plans, these legal sanctions would probably be
inadequate. The proposed legal sanctions involve withholding of
construction grants and/or permits in the absence of complete
planning. Withholding a construction grant is inadequate because it
would give a municipal discharger a reason for not building a sewage
treatment plant. Even with a 75 percent federal subsidy for construction costs, a municipality must cover approximately 50 percent

of the annual costs (25 percent of the capital costs and all of the
operation and maintenance costs). In addition, the history of the
act to date suggests that such action would be politically unfeasible.
Similarly, withholding a permit from either a municipal or industrial discharger is inadequate because it would only create the basis
for more delays in the courts rather than positive steps toward
restoring the nation's waters. Nor would citizen suits have much
impact on industrial dischargers, if the current track record is any
indication of their success in insuring the actual installation and
effective operation of water pollution control technology where it is
needed to prevent environmental deterioration (Environmental Law
Institute 1975).

Furthermore, areawide plans and programs completed by
the late 1970s would be timely rather than too late. Both the EPA
and NCWQ reports suggest that the 1977 objective would not be met
until the early 1980s with some possibility of meeting specific
industrial requirements at an earlier date (General Accounting
Office 1974, and National Commission on Water Quality 1976). If
so, completion of areawide plans in the late 1970s would be timely.
They would be available if each area were given, following our
recommendation, the choice of (1) justifying not installing new
additional control technology, (2) indicating necessary source-
specific, cost-effective solutions, or (3) planning to implement
BAT/BPWTT requirements for point sources as well as additional
areal-source control measures where needed.

Possible Legislative Amendments

First, Congress should modify only slightly the 1977 deadline for installing BPT/ST. Some delay in the deadline is necessary
to maintain the credibility of the water quality program. However,
a delay of the 1983 deadline is inadvisable. A delay in this target
date would eliminate the principal reason for point-source dischargers
to seriously participate in areawide planning because they would
no longer face an immediate investment requirement.

Second, Congress should reinforce the pivotal role of areawide plans and programs. Congress ought to allow municipal and
industrial dischargers to be exempted from BAT/BPWTT if areawide plans support alternative courses of action.

Third, Congress should reaffirm the central role of comprehensive areawide planning. Congress ought to encourage EPA to
integrate its numerous and often fragmented planning activities
into an effective action program.

Fourth, Congress should modify the procedure for completing areawide plans in order that they become documents capable of justifying, if needed, a midcourse correction. The present regulations for areawide planning do not allow financing of lead time necessary to accomplish assembling of staff, inventorying of existing data availability, and the programming of the development of needed data for planning (Harold Wise and Associates. 1975). Congress ought to insure that start-up money is available and that areawide planning agencies have a full two years to complete their work.

Practicality of Areawide Plans

While no one can definitely answer whether areawide plans could become a successful basis for a midcourse correction, there are both encouraging and discouraging signs.

The federal and state governments are making a large financial commitment to areawide planning and its proposed complement, basin planning. The federal government has committed itself to $300 million (100 percent federal financing) for areawide planning agencies. While most of these agencies are in SMSAs, some are in rural areas. Commitment of an additional $60 million of federal and state funds for basin planning could accomplish the same ends in geographic areas not covered by areawide plans and programs.

However, the performance of the federal and state agencies to date in implementing the permit program makes one very pessimistic about assembling data even on point-source dischargers in a reasonable time. While EPA and state counterparts have issued 98 percent of the major and 59 percent of the minor industrial permits, data from these agencies are not available for national analysis of the major industries studied in depth by NRDI. Data on production, residuals discharge, or in-place technology are not available for most of the permits nor are the permits computerized for availability for national analysis. If EPA and state agencies could not obtain minimal information about major dischargers by 1976, local and regional authorities might have difficulty meeting this objective.

Moreover, sufficient skilled personnel apparently are not available to EPA or state areawide planning agencies. While this claim is difficult to document quantitatively, a review of the initial decision of many of the areawide planning authorities leaves one very skeptical about whether they can assemble and interpret data sufficiently well to formulate a midcourse correction.

In spite of the potential for failure, Congress must seriously consider this recommendation. If the 1972 Law is not amended as proposed by the EPA, then the nation may invest its capital resources needlessly and may fail, in spite of its effort, to restore water quality. If the 1972 Law is changed too much, primarily by following the NCWQ recommendation to shift the BAT/BPWTT objective so far forward in time as to make it meaningless, then it would lose its force to make point-source dischargers participate meaningfully in area planning. Both the potential for overinvesting capital and for eliminating the threat of additional capital investment could be avoided by amending the law only slightly to allow areawide plans to become the basis of the midcourse corrections.

This recommendation also merits serious consideration in light of the current emphasis on decentralization of government decision making. Placing the final decision about the midcourse corrections with local authorities is consistent with an attempt to return decisions to local authorities. Either retaining the law as is or delaying the 1983 technological objective for five to ten years for the nation as a whole would be inconsistent with this philosophy (Lieber and Rosenoff 1975). Both alternatives leave the focus of decision making in the federal government rather than giving local citizens an opportunity to affect their own environment.

Finally, the recommendation merits serious consideration in light of the realities of water quality management. It recognizes the technocrats' limited ability to understand water quality management and to provide sufficient data to national decision makers. For all the monies invested by EPA and NCWQ, methodologies are not available to provide the linkage between effluent limits for selected residuals, water quality changes associated with those residuals, and changes in beneficial use as a function of the changes in ambient levels of selected residuals. A similar concern has been voiced by Ackerman et al. (1974). A major defect in the law at present is its failure to recognize that local wisdom and regional preferences are as important to an overall assessment as are the limited data on benefits and damages available to national decision makers. The former inputs are most likely to be influential if local authorities play a significant role in the assessment of future plans.

REFERENCES

Akerman, B. et al. 1974 The Uncertain Search for Environmental Quality. New York: Free Press.

Environmental Law Institute. 1975. Water Pollution Control Act of 1972, Institutional Assessment—Enforcement. PB 246 320. Springfield, Va. : National Technical Information Service.

Energy and Environmental Analysis. 1975. Water Pollution Control Act of 1972, Institutional Assessment—The Permit Program. PB 244 805. Springfield, Va.: National Technical Information Service.

Federal Register. 1975. Volume 40, No. 230, November.

General Accounting Office. 1974. Implementation of Federal Water Pollution Control Act Amendments of 1972 Is Slow. Washington, D.C.: General Accounting Office.

Harold Wise and Associates. 1975. Water Pollution Control Act of 1972, Institutional Assessment—Planning. PB 244 907. Springfield, Va.: National Technical Information Service.

Lieber, H., and Rosenoff, B. 1975. Federalism and Clean Waters. Lexington, Mass.: Lexington Books, D.C. Heath and Co.

National Commission on Water Quality. 1976. Staff Report to the National Commission on Water Quality. Washington, D.C.: Government Printing Office.

SUMMARY OF NCWQ FINDINGS
AND RECOMMENDATIONS

The NCWQ Staff Report (National Commission on Water Quality 1976a) does not contain an executive summary of the five major study areas. Thus the following summary is based on the National Research Council's Review of the Staff Draft Report (1976).

TECHNOLOGY FOR AND COST OF WATER QUALITY IMPROVEMENTS

The NCWQ report concludes that, by and large, technology for both municipal and industrial dischargers is available to meet the 1977 and 1983 requirements. However, the report makes clear that some EPA effluent requirements or guidelines for 1983 for specific industries entail the use of technologies not yet used in prototype conditions in those industries. In addition, even in relation to the 1977 requirement, temporal variations in the generation of effluents, and in their quality and quantity, pose problems in meeting short-time or peak effluent limitations. Compliance with effluent limitations for short time periods as contrasted with compliance with a monthly mean becomes increasingly difficult as more stringent limitations are applied.

Based on 1990 conditions, the capital costs of meeting the 1983 requirements for treatment works for municipalities are estimated by NCWQ to be $10.8 billion for secondary treatment and $24.8 billion to meet more stringent levels of water quality for limited reaches, for a total of $35.6 billion. Estimated costs for new interceptors are $13.5 billion. These represent the major categories for which federal funds are being expended. (The amounts represent investment in addition to existing facilities.) Additional annual operation and maintenance costs are estimated to be $1.5 billion, to meet the 1977 requirements.

While neither industrial dischargers nor municipalities will meet the 1977 requirements by the July 1, 1977 deadline, it appears that industry will achieve the requirement earlier than publicly owned treatment works.

Based on 1990 conditions, NCWQ estimates of the cost of correcting combined sewer overflows, depending on control and treatment technology, range from $5.4 billion to $88 billion. Control and treatment of storm sewer discharges in major urbanized areas range from $60 billion to $454 billion depending upon the storm events criteria used and the level and method of control.

Based on 1973 conditions, the capital costs of meeting the 1977 requirements for industrial discharges are estimated to be $44.3 billion. Additional operations and maintenance costs, over and above present costs, are estimated to be $7.7 billion annually, to meet the 1977 requirements.

Based on 1973 conditions, the incremental capital costs of moving from the 1977 to the 1983 requirements for industrial point discharges are estimated to be $30.6 billion. Corresponding additional operations and maintenance costs are estimated to be $6.2 billion annually.

MACROECONOMIC AND SOCIAL EFFECTS

NCWQ reports that the estimated municipal and industrial expenditures to meet the 1977 and 1983 requirements represent 0.5 percent to 1.5 percent of forecasted GNP.

Disaggregating at the regional level, the estimated municipal and industrial expenditures to meet the 1977 and 1983 requirements will impact differentially upon different regions, with the hardest hit being those, such as New England, with older plants and a less robust economy.

For the industries studied in detail the average cumulative price increase by 1985 due to the BPT requirements is estimated to be 4.1 percent—for most industries the price increase is less than 2 percent, and for one, 35 percent. Additional increases in price associated with meeting BAT requirements are estimated to be less than 1 percent except in the metal finishing industry where the estimated increase is 43 percent.

Plant closings as a result of expenditures to reduce discharges will occur primarily among plants already in a marginal economic position. A number of these plants are in the Northeast, an area including economically depressed communities often dependent on these marginal enterprises.

Monetary benefits from improving water quality were based on four categories of impacts: marine commercial and sports harvesting of shellfish and finfish, reopening of beaches for swimming presently closed to swimming for water quality reasons, freshwater recreational activities, and selected types of water-oriented residential property. The estimated annual rate of these benefits is $3.4 billion by 1980, $5.2 billion by 1985, and $7.8 billion by 2000.

IMPACT OF IMPLEMENTATION OF
PL 92-500 ON WATER QUALITY

At the sample sites studied, in terms of customary measures of pollutants such as oxygen depletion, nutrient enrichment, and high bacteria counts, the most severely polluted reaches of rivers, lakes, impoundments, estuaries, and coastal regions are those dominated by large human populations and major industrial activities. In such areas where publicly owned treatment works and industrial activities discharge materials from point sources to adjacent waters, at the sample sites studied, attainment of the 1977 requirements will result in significant improvement in water quality as measured by levels of dissolved oxygen. Meeting 1983 requirements is likely to produce a significantly smaller incremental improvement in water quality measured in terms of dissolved oxygen.

The elimination of discharge of materials and waste heat from point sources results in still smaller incremental improvements at many of these same sites. At some others it might be hypothesized that reduction in discharges of substances other than oxygen-demanding wastes, such as some potential toxic substances, beyond that associated with standard biological treatment processes, could result in water quality improvement, but these substances have not been measured, nor their effects estimated.

At a number of sites, it is reasonable to expect that the size and/or composition of the fish population will be improved by reduction in discharges from point sources to meet the 1977 requirements. At some, shellfish habitats may be similarly improved. More modest improvements are projected to accompany achievement of the 1983 requirements. Improvements in fish habitats (and presumably populations) at these latter levels appear to be projected primarily for streams characterizing the Middle Atlantic, Northeast, and Midwest regions. Fish kills are projected to decrease at several sites with progressive reduction of discharges from point sources.

Nonpoint sources of pollution, such as urban and agricultural runoff, presently account for a large proportion of the materials discharged to water bodies at some locations. The relative magnitude of nonpoint-source discharges of specific substances varies greatly at different locations and in different regions of the country. Rivers in the Midwest and in the Southwest carry large quantities of suspended solids. Similarly, some also contain high concentrations of dissolved solids. In many regions runoff from agricultural areas may also contain high concentrations of nutrients, phosphorus, and nitrogen, along with oxygen-demanding materials and pesticides. For many large urban and industrial centers—for example, Houston and Atlanta—the percentage of total discharges represented by discharges from nonpoint sources, urban runoff, and combined sewer systems, will increase as discharges from point sources decrease.

INSTITUTIONS INVOLVED IN IMPLEMENTING PL 92-500

The tight schedule mandated by the act for the issuance of rules, regulations, effluent guidelines and limitations, and permits, combined with delays in funding and the necessity of developing administrative procedures for the large construction grants program have overtaxed the administrative capacity of EPA. BPT for industry and for publicly owned treatment works will not be achieved by the 1977 deadline specified in the act. Industry is expected to achieve BPT by 1980. Achievement of BPT by publicly owned treatment plants is dependent upon the rate of federal funding.

Promulgation of effluent guidelines has proved to be both difficult and a source of delay. Both complexity and delay are evident in the fact that effluent guidelines are being contested by industry in more than 250 legal actions. Consolidated in the U.S. Court of Appeals to 21, these judicial actions include 21 of the 42 industrial categories for which 1983 effluent guidelines have been promulgated.

Permits under the National Pollutant Discharge Elimination System cannot be made identical to uniform effluent limitations because of the variety of industrial activities and the number and combination of pollutants discharged by individual plants, and water quality conditions at the individual sites.

Slow progress in initiating obligation of federal construction grant funds has delayed achievement of 1977 requirements by publicly owned treatment works. At current rates the $18 billion designed by Congress would not be spent until 1982, five years or more after the mandated date. Delays are attributed to normal problems of start-up coupled with excessive red tape, duplication of reviews, inadequate staffing, and overloading of available consulting firms needed to plan and design facilities.

Compliance monitoring of dischargers operating under permits has, to date, been inadequate to serve as a mechanism for enforcement or as a management tool in monitoring performance.

Thus far, little or no effective planning has been implemented under PL 92-500. Construction grants and permit conditions have led, rather than have been directed by, water quality management planning.

Implementation of PL 92-500 has proven to be more complex than framers of the legislation anticipated. The act required

federal, state, and local governments, dischargers, and citizens
to play roles in implementing the many requirements and procedures set forth in the act. Initially these participants have often
become adversaries rather than cooperators.

NCWQ RECOMMENDATIONS SUMMARY

The NCWQ Report to Congress (1976b) lists the midcourse
corrections that NCWQ thinks would keep the nation's water pollution control program on an effective course toward the objectives
of the act. The recommendations are as follows:

1. Maintain the July 1, 1977 date for compliance with uniform
treatment requirements by both industry and publicly owned treatment works, but provide some flexibility to grant extensions, and
even waivers, on a case-by-case and category-by-category basis.

2(a). Maintain the 1983 interim water quality goal, while postponing the 1983 requirements for application of uniform technologies
for five to ten years. pending an assessment of progress in water
quality improvement and review of these results by a new NCWQ
by 1985; (b) meet the 1983 interim water quality goal through application of the 1977 requirements to all dischargers; revisions of
1977 limitations; effluent limitations for the elimination of the discharge of toxic pollutants in toxic amounts, beginning immediately;
new source performance standards for all new point-source discharges; periodic upgrading of permits for discharges into limited
waters; and application of control measures to combined sewer
overflows, urban stormwater runoff, agricultural and nonpoint
sources.

3. Decentralize regulatory and administrative functions of the
national program by selective certification of states, based on
satisfactory state plans and programs to control both point and
nonpoint sources (including irrigated agriculture).

4. Stabilize the federal construction grants program by assuring
75 percent federal financing for priority treatment needs at a fixed
amount (not less than $5 billion nor more than $10 billion per year)
for a specified number of years (five to ten).

5. Redefine the goal of elimination of discharge of pollutants
as one stressing conservation and reuse of resources.

6. Authorize flexibility in applying control or treatment measures to irrigated agriculture after an inventory of the problem,

and support salinity alleviation projects to reduce salt loads from
sources other than man's activities.

Specific recommendations from the NCWQ Report to Congress
relative to the 1977 and 1983 effluent limitations are outlined in
detail below.

THE 1977 REQUIREMENTS

I.   The Commission Recommends That:
    A. Congress authorize granting extensions of time to munici-
        pal, industrial and agricultural dischargers to meet the
        1977 requirements on a case-by-case basis where:
        1. The discharger can demonstrate reasonable progress
            toward compliance with the July 1, 1977 deadline; or
        2. Lack of Federal construction grant funds has caused
            delay; or
        3. The discharger can demonstrate other good and
            sufficient cause.
        Provided that in no case shall such extensions of time
        extend beyond (a specified date such as September 30,
        1980) or until the cause for delay has been removed.
    B. Congress authorize waiving, deferral or modification
        of the 1977 requirements on a case-by-case basis where
        the discharger demonstrates to the satisfaction of the
        Administrator (or a state administrator where a state
        has been certified) that adverse environmental impacts
        of such action will be minimal or nonexistent, or that
        the capital or operation and maintenance costs are dis-
        proportionate to projected environmental gains.
    C. Congress authorize waiving, deferral, or modification
        of the 1977 requirements on a category-by-category basis
        for near shore ocean discharges of publicly owned treat-
        ment works, pretreatment requirements, existing pub-
        licly owned treatment works, pretreatment requirements,
        existing publicly owned oxidation ponds and lagoons, and
        deminimus situations where the Administrator determines
        that the adverse environmental impacts of such action will
        be minimal or nonexistent, or that the capital or operating
        and maintenance costs are disproportionate to projected
        environmental gains.

THE 1983 GOAL AND REQUIREMENTS

II.   THE COMMISSION RECOMMENDS THAT:

A.   Congress retain the national goal, ". . . that wherever attainable, an interim goal of water quality which provides for the protection and propagation of fish, shellfish, and wildlife and provides for recreation in and on the water be achieved by July 1, 1983."

B.   Congress postpone the deadline by which municipal, agricultural, and industrial dischargers shall be required to meet the 1983 requirements from July 1, 1983 to (a date not less than five and no more than ten years after 1983) provided the following interim actions are assured:

1. Effluent limitations for 1977 are reviewed periodically and revised, if appropriate, to reflect advances in practicable control technology.

2. Periodically review and aggressively enforce higher levels of treatment than required by the 1977 effluent limitations where the 1977 requirements will not achieve Federally approved state water quality standards and where more stringent limitations will significantly help in achieving water quality standards.

3. Review and alter new source performance standards periodically as technology is perfected to ensure a high level of control or treatment of new pollutant sources.

4. a. Where possible, toxic pollutants in toxic concentrations shall be controlled in applicable effluent limitations and permits.

    b. Effluent limitations based on technology to eliminate the discharge of toxic pollutants in toxic concentrations into the Nation's waters shall be implemented as soon as possible but no later than October 1, 1980.

5. a. Apply control or treatment measures to combined storm and sanitary sewer flows and to urban stormwater flows when these measures are cost effective and will significantly help in achieving water quality standards.

    b. Control or treatment measures shall be applied to agricultural and nonpoint discharges when these measures are cost effective and will significantly help in achieving water quality standards. For these measures, Congress should utilize the capabilities of existing institutions and their resources, and may wish to consider additional Federal resources to carry out the necessary programs.

6. An on-going national assessment of the quality of the Nation's waters shall be undertaken to determine progress toward water quality goals and objectives, and the progress periodically reported to the Congress.

7. No later than 1985 a Commission similar to the National Commission on Water Quality shall evaluate progress toward these goals and make appropriate recommendations, at which time Congress may consider whether uniform application of more stringent effluent limitations than the 1977 requirements is justified and desirable.

REFERENCES

National Commission on Water Quality. 1976a. Staff Report to the National Commission on Water Quality. Washington, D.C.: Government Printing Office.

_____. 1976b. Report to the Congress of the National Commission on Water Quality. Washington, D.C.: Government Printing Office.

National Research Council. 1976. November 1975 Staff Draft Report, Review and Comment. Washington, D.C.: National Research Council.

Region/River Basin	Basin Number	1975 Estimated Population (in thousands)	Total Area In Acres (in thousands)
New England			
St. Johns-St. Croix, Penobscot, Kennebec-Androscoggin	101	667	18,624
Saco, Merrimack	102	921	3,904
Marrmode Is Coastal	103	6,153	3,904
Long Island Sound	104	2,380	3,008
Connecticut	105	1,999	7,808
St. Francis, Richelieu	106	372	4,800
Total		12,492	42,048
Middle Atlantic			
Upper Hudson	201	2,145	9,472
Lower Hudson	202	17,023	2,816
Delaware	203	7,982	9,664
Susquehanna	204	3,676	17,152
Upper Chesapeake, Lower Chesapeake	205	4,673	15,360
Potomac	206	4,113	9,472
Total		39,612	63,936
South Atlantic			
Roanoke, Tar-Reuse, Cape Fear	301	3,133	22,784
Pee Dee, Sante-Edisto	302	4,939	26,432
Savannah-Ogeechee, Atlamana-St. Mary's	303	1,973	22,400
St. Johns, Tampa, Bay, Swannee	304	3,979	17,600
Southern Florida	305	3,486	11,264
Ochlockone, Apalachicola	306	2,960	17,984
St. Josephs-Perdido, Alabama	307	1,890	20,288
Tombigbee	308	1,901	15,168
Pascogaula, Pearl	309	1,162	12,800
Total		25,423	166,720

Great Lakes

W Superior, S Superior	401	535	16,256
NW Michigan	402	1,077	10,496
SW Michigan	403	9,988	6,208
SE Michigan, NE Michigan	404	3,110	16,512
NW Huron, SW Huron	405	1,316	8,448
St. Clair-Detroit, West Erie	406	6,929	10,304
S Erie, E Erie	407	5,056	5,376
SW-SE-NE Ontario	408	2,381	10,688
Total		30,392	84,288

Ohio

Allegheny, Monongahela	501	1,342	9,664
Pittsburgh-Wheeling, Portsmouth-Little Knanwha, Cincinnati-Little Miami	502	6,664	17,728
Musgnkm, Sciot, Great Miami	503	4,165	12,928
Kanawha	504	895	8,256
Lickin-Ky, Louisville-Salt, Evansville-Green	505	2,960	20,032
White-Patoka, Wabash	506	3,739	21,504
Cumberland	507	1,393	11,072
Total		21,158	101,184

Tennessee

Upper Tennessee-Tennessee-Hiwassee-Sequatchie	601	2,467	15,360
Tennessee-Elk, Lower Tennessee	602	1,098	11,200
Total		3,565	26,560

Upper Mississippi

MN, MS Headwaters, St. Croix	701	3,035	27,008
MS-Black-Root, Wisconsin	702	1,244	18,624
MS-Maquoketa-Plum, Rock, Des Moines, Iowa Quad	703	4,009	35,392
MS-Salt-Quincy, Upper Illinois, Lower Illinois	704	1,961	19,392
MS-Kaskaskis-St. Louis	705	3,138	11,840
Total		13,387	112,256

Lower Mississippi

MS-Hatchie, St. Francis	801	1,962	17,920
MS-Yzao, Quachta, Tenas, Big Black	802	1,849	31,040

MS-Lake Maureps, La Coast, Delta	803	2,606	15,552
Total		6,417	64,512
Souris-Red-Rainy	901	649	33,920
Missouri			
Milk-Me-Poplar	1001	61	16,896
Mo Headwaters, Mo-Marias	1002	200	23,168
Mo-Musselshell	1003	22	10,752
Upper Yellowstone, Big Horn, Tongue-Powder, Lower Yellowstone	1004	281	47,630
Mo-Little Mo, Cheyenne, Oake, Mo-White	1005	420	59,840
James, Mo-Big Sioux	1006	590	23,552
N Platte, S Platte	1007	1,968	38,080
Niobrar, Loop, Platte, Elkhorn	1008	622	46,268
Mo-Sioux City-Omaha, No-Nemaha-Nodaway	1009	1,224	22,032
Republican, Smoky Hill, Kansas	1010	901	54,768
Grand, Chariin, Osage-Gascnade, Mo-Kansas City	1011	2,543	40,311
Total		8,832	388,297
Arkansas-White-Red			
White	1101	352	12,544
Upper Arkansas	1102	457	15,872
Ar-Ks, Upper Cimarron, Lower Cimarron, Keystone	1103	970	29,504
Verdigrs-Nedshd, Lower Arkansas	1104	2,086	25,024
Upper Canadian, Tx Canadian, Lower Canadian	1105	1,088	46,247
Red Headwaters, Red-Washita	1106	926	39,579
Lower Red	1107	967	16,064
Total		6,846	184,834
Texas Gulf			
Sabine, Neches	1201	967	10,880
Upper Trinity, Low Trinity	1202	5,050	16,704

Brazos Headwaters, Middle and Low Brazos	1203	1,282	46,263
Colorado Headwaters, Lower Colorado Land	1204	1,017	46,394
Guadalue-San Antonio, Nueces-Frio	1205	1,595	24,192
Total		9,911	144,433
Rio Grande			
Headwaters	1301	96	5,120
North Rio Grande- Mimbres, Closed Basins	1302	1,004	37,440
Big Bend, Low Pecos	1303	71	21,248
Upper Pecos	1304	110	13,056
Amistad, Low Rio Grande	1305	474	10,176
Total		1,695	87,040
Upper Colorado			
Green, Yampa-White, Lower Green	1401	94	29,504
Gunison, Co Headwaters, Colorado-Dolores	1402	141	16,384
Upper San Juan,	1403	108	19,520
Co-San Juan		343	65,408
Total			
Lower Colorado			
Little Colorado	1501	127	16,960
Lake Mead-Lake Mohave	1502	474	40,192
Upper Gila, Gila-San Pedro-Gila-Salt	1503	1,811	41,600
Total		2,412	98,752
Great Basin			
Bear, Great Salt Lake	1601	975	15,936
Sevier Lake	1602	48	12,982
Humboldt, Tonopah Desert	1603	43	46,784
Central Lahontan	1604	196	11,520
Total		1,262	87,232
Columbia-North Pacific			
Pend Oreille, Kootenai, Spokane	1701	610	22,848
Yakima, Upper Columbia, Deschutes, Middle Columbia	1702	665	38,016
Upper Sanke, Middle Snake	1703	549	41,600
Salmon, Lower Snake	1704	171	20,288

Willamette, Lower Columbia, Washington- Oregon Coastal	1705	2,306	24,576
Puget Sound	1706	2,389	10,048
Oregon Closed Basin	1707	14	11,776
Total		6,704	169,152
California-South Pacific			
North Coastal	1801	266	15,040
Sacramento Basin	1802	1,313	20,224
Tular Bain, San Joaquin, Delta Central Sierra	1803	1,714	20,864
San Francisco Bay	1804	4,988	4,416
Central Coastal	1805	800	7,168
S Coastal, CO Desert	1806	12,056	27,200
S Lahontan	1807	23	8,832
Total		21,160	103,744

*Coastal Waters and bays omitted.

Note: Abbreviations are those authorized by the Water Resources Council.

Source: Water Resource Council. 1970. Water Resources Regions and Subregions for the National Assessment of Water and Related Land Resources. Washington, D.C.: Water Resources Council; and Water Resource Council. 1974. 1972 OBERS Projections. Washington, D.C.: Government Printing Office.

REFERENCES

Water Resources Council. 1970. Water Resource Regions and Sub-regions for the National Assessment of Water and Related Land Resources. Washington, D.C.: Water Resources Council.
_____. 1974. 1972 OBERS Projections. Washington, D.C.: Government Printing Office.

APPENDIX C
UNIT PROCESS DEFINITIONS,
COSTS, REMOVALS, AND
COMPENSATION FACTORS

This appendix defines each of the 39 unit processes used in the point-source analyses to compute residuals reductions and to estimate capital costs.

The use of a single set of unit processes in the municipal, special industry, and general industry analyses provides a consistent framework for both cost and residual analyses, which cut across categories.

This approach used in the NRDI is quite different from that used by both EPA and NCWQ, in which each contractor analyzing an industry group was responsible for independently developing cost estimates. While many of these contractors used certain unit processes similar to those described here, in most cases, documentation of the derivation of the cost functions was inadequate to permit reconstruction of the estimates reported. Thus, although the EPA/NCWQ approach has the theoretical advantage of providing for industry-to-industry differences in certain similar treatment processes, this advantage is outweighed by the difficulties of relating residuals to costs and understanding derivation of costs.

The NRDI approach was designed with the objectives of providing an easily understood and consistent set of unit processes for building national and regional cost estimates; and relating treatment scheme performance to the specific documented residual removal efficiencies of the individual unit processes.

Basic information concerning the 39 unit processes used is presented in Table C.1. Although six different sources were used for cost function derivation, the majority of the data (27 functions) came from the Metcalf and Eddy (1975) report to NCWQ. The other functions were derived from other sources since they were processes not used for municipal treatment and thus not reported in the Metcalf and Eddy study.

The use of identical cost functions for municipal and industrial treatment is unusual since it is commonly assumed that industrial facilities are designed for a shorter service life and are thus less costly to construct. However, a review of the literature on costs of industrial waste treatment processes showed that many of the sources for cost estimates were reports on municipal costs. Thus, it was thought that the use of common cost functions was justified.

TABLE C.1

Characteristics of NRDI Cost Functions and Unit Processes

Unit Process	Process Abbreviation	Influent Concentration used in Computing Costs			Purpose of Process
		BOD	SS	COD	
Activated sludge	AC	X			Treatment
Aerated lagoon	AE	X			Treatment
Trickling filter	TR				Treatment
Neutralization	NE				Treatment
Flotation	FL				Treatment
Petroleum flotation	PE				Treatment
Cooling towers	CO				Thermal
Recirculation	RE				Flow reduction
Surface outfall	SU				Liquid disposal
Ocean outfall	OC				Liquid disposal
Extended biological oxidation	XA				Treatment
Chlorination	CL				Treatment
Earthen basin	EA				Treatment
Chemical addition	CH				Treatment
Primary sedimentation	PR				Treatment
Filtration (multimedia)	MU				Treatment
Oil separation	OI				Treatment
Spray irrigation	SP				Treatment*
Equalization	EQ				Treatment
Nitrification	NI				Treatment
Ammonia stripping	AS				Treatment
Nitrification-denitrification	AS				Treatment
Drying beds	ND				Treatment
Incineration	DR	X	X		Sludge handling
Landfill	IN	X	X		Sludge handling
Vacuum filtration	VA	X	X		Sludge handling
Gravity thickening	GR	X	X		Sludge handling
Heat treatment	HE	X	X		Sludge handling
Sludge digestion	SL	X	X		Sludge handling
Chrome reduction	CR				Treatment
Sedimentation with chemical addition	SI				Treatment
Carbon adsorption	CA				Treatment
Breakpoint chlorination	BR				Treatment
Ion exchange	IO				Treatment
Two-stage chemical addition	TW				Treatment
Evaporation	EV				Treatment*
Holding pond	HO				Treatment
Cyanide destruction	CY				Treatment
Screening	SC				Treatment

Note: All functions use wastewater flow.

*Also includes liquid disposal.

Sources: Associated Air and Water Resources Engineers. 1973. Estimating Water Pollution Control Costs from Selected Manufacturing Industries in the U.S. 1973-1977. Washington, D.C.: Environmental Protection Agency; Datagraphics Inc. 1975. Wastewater Treatment Process Cost Function. Carnegie, Pa., mineographed; Environmental Protection Agency. 1973. Process Design Manual for Carbon Adsorption. Washington, D.C.: Environmental Protection Agency; Environmental Protection Agency. 1975a. Development Document for Final Effluent Limitations for Fruits, Vegetables, and Specialties. Washington, D.C.: Government Printing Office; Environmental Protection Agency. 1975b. Development Document for Interim, Final, and Proposed Effluent Limitation Guidelines for Ore Mining and Dressing Industries. Vol. 2. Washington, D.C.: Government Printing Office; and Metcalf and Eddy, Inc. 1975. Report to the National Commission on Water Quality: Assessment of Technologies and Costs for Publicly Owned Treatment Works Under P.L. 92-500. PB 250 690. Springfield, Va.: National Technical Information Service.

CAPITAL COST ESTIMATION

Facilities were sized and costs were estimated based on the raw influent characteristics of flow, BOD concentrations, SS concentration, and COD concentration. Flow was used in all cost analyses. BOD and SS were used with flow in sizing sludge handling facilities, BOD was used with flow in sizing activated sludge and aerated lagoons, and COD was used with flow in sizing activated carbon adsorption processes.

Except for carbon adsorption, all cost functions were of the basic form

$$C = aQ^b$$

where C is construction costs (in millions of 1973 dollars), Q is adjusted flow (in millions of gallons per day) or sludge loading (in thousands of pounds per day), and a and b are constants defining the function. Since a and b are not constant over all ranges of Q, piecewise continuous functional forms were used. Slightly under 100 of these piecewise continuous segments were used for 38 functions.

Carbon adsorption was a special case in that the unit process cost consisted of a number of separate entities. For that reason, a special cost routine was prepared, which produced a special estimate for each facility.

The cost functions are presented in tabular form for a variety of plant sizes and a set of influent loading assumptions. This is done in Table C.2 for seven plant sizes and a municipal strength influent.

Construction costs were converted to capital costs by application of adjustment factors. This procedure is discussed in a later section.

REMOVAL EFFICIENCES

Each treatment process (Table C.1) was characterized by residual removal rates. BOD and SS removals were computed for all processes and P and N removal were computed for functions used in the municipal model. The results of the estimates are presented in Table C.3.

These removal rates were used exclusively in the municipal and general industry models. In the special industry model, certain of the removal efficiencies were modified based on detailed knowledge of the waste characteristics and how they are affected by the various processes. An example is the iron and steel industry, which achieved greater-than-indicated SS removals with sedimentation due to the heavy particles. These modifications are described in earlier chapters.

TABLE C.2

Construction Costs for Unit Processes, by Plant Size
(millions of 1973 dollars)

Process	Plant Size (millions of gallons per day)						
	.1	1.0	5.0	10.0	20.0	50.0	100.0
Activated sludge	0.150	0.630	1.700	2.629	4.067	7.240	11.199
Aerated lagoon	0.090	0.393	1.100	1.714	2.671	4.801	7.481
Trickling filter	0.051	0.220	0.984	1.876	3.576	8.391	15.996
Neutralization	0.022	0.088	0.231	0.350	0.532	0.924	1.402
Flotation	0.120	0.200	0.420	0.699	1.163	2.282	3.799
Petroleum flotation	0.020	0.111	0.361	0.601	1.000	1.961	3.265
Cooling towers	0.010	0.042	0.110	0.166	0.252	0.436	0.661
Recirculation	0.106	0.421	1.106	1.676	2.540	4.402	6.672
Surface outfall	0.010	0.046	0.136	0.215	0.342	0.630	1.000
Ocean outfall	0.100	0.531	1.708	2.823	4.668	9.073	15.002
Extended biological oxidation	0.053	0.190	0.490	0.737	1.108	1.900	2.857
Chlorination	0.020	0.076	0.194	0.289	0.433	0.736	1.100
Earthen basin	0.002	0.011	0.033	0.063	0.123	0.227	0.361
Chemical addition	0.009	0.025	0.044	0.080	0.147	0.328	0.600
Primary sedi- mentation	0.120	0.200	0.420	0.699	1.163	2.282	3.799
Filtration (multi- media)	0.100	0.200	0.570	0.894	1.404	2.548	4.000
Oil separation	0.013	0.087	0.335	0.601	1.075	2.321	4.155
Spray irrigation	0.062	0.210	0.680	1.128	1.871	3.653	6.059
Equalization	0.100	0.150	0.313	0.430	0.683	1.260	2.001
Nitrification	0.230	0.400	1.313	2.191	3.656	7.193	12.002
Ammonia stripping	0.243	0.310	0.966	1.575	2.569	4.905	8.001
Nitrification- denitrification	0.510	0.800	2.527	4.148	6.807	13.102	21.503
Drying beds	0.010	0.042	0.153	0.301	0.592	1.447	2.844
Incineration	0.144	0.719	2.212	2.500	2.540	4.119	5.768
Landfill	0.013	0.063	0.192	0.310	0.500	0.941	1.519
Vacuum filtration	0.099	0.313	0.697	0.864	0.896	0.941	1.557
Gravity thickening	0.014	0.046	0.106	0.159	0.254	0.473	0.758
Heat treatment	0.128	0.323	0.620	0.820	1.085	1.887	3.696
Sludge digestion	0.075	0.298	0.787	1.231	2.008	3.833	6.253
Chrome reduction	0.515	1.638	3.663	5.180	7.325	11.582	16.380
Sedimentation with chemical addition	0.129	0.225	0.464	0.781	1.314	2.615	4.401
Carbon adsorption	0.357	0.965	2.030	3.230	5.392	11.157	20.016
Breakpoint chlori- nation	0.055	0.185	0.433	0.624	0.899	1.457	2.100
Ion exchange	0.232	0.450	1.366	2.205	3.557	6.694	10.800
Two-stage chemical addition	0.160	0.450	1.246	1.932	2.997	5.352	8.299
Evaporation	0.412	2.028	6.180	9.985	16.135	30.427	49.166
Holding pond	0.250	0.480	1.200	1.990	3.300	6.440	10.679
Cyanide destruction	0.515	1.638	3.663	5.180	7.325	11.582	16.380
Screening	0.015	0.028	0.076	0.116	0.178	0.313	0.480

Note: Municipal influent loading assumptions: BOD = 200; SS = 200; and COD = 500 milligrams per liter.

Source: NRDI Computer Outputs. 1975.

TABLE C.3

Removal Efficiencies of Unit Processes
(percent)

Function	Removal Efficiency[a]			
	BOD[b]	SS	P	N
Activated sludge	85	77	22	20
Aerated lagoon	75	65	10	20
Trickling filter	80	68	10	20
Neutralization	0	0		
Flotation	40	80		
Petroleum flotation	40	80		
Extended biological oxidation	75	65		
Chlorination	0	0		
Earthen basin	0[c]	80		
Chemical addition to earthen basin	0[c]	80		
Primary sedimentation	33	52	9	10
Filtration (multimedia)	50	72	50	0
Oil separation	0	0		
Spray irrigation	100	100		
Equalization	0	0		
Nitrification[d]	70	57	0	0
Ammonia stripping	0	0		
Nitrification-denitrification	0	0	0	90
Chrome reduction	0	0		
Sedimentation with chemical control	52	71	94	0
Carbon adsorption[e]	60	60	20	0
Breakpoint chlorination	0	0		
Ion exchange	0	0		
Two-stage chemical addition	60	68		
Evaporation	100	100		
Holding pond	79	68	10	0
Cyanide destruction	0	0		
Screening	0	0		

[a]P and N removals shown only for municipal functions.

[b]For municipal, BOD removals for primary and secondary combined are limited to 85 percent.

[c]Earthen basins used only for wastes with zero BOD loadings.

[d]Municipal only.

[e]BOD and SS were, respectively, 50 and 72 percent for municipal.

Source: Bechtel, Inc. 1975. Cost Effective Wastewater Treatment Systems. Washington, D.C.: Environmental Protection Agency.

TABLE C.4

Unit Process Technology Costs, by Flow
(millions of 1973 dollars)

Sequence of Unit Process Technologies	Flow (millions of gallons per day)			
	.1	1.0	5.0	10.0
Primary sedimentation	.117	.220	.420	.707
Activated sludge	.149	.630	1.700	2.640
Gravity thickening	.012	.061	.184	.298
Vacuum filtration	.099	.312	.695	.864
Landfill	.015	.049	.112	.146
Total	.392	1.252	3.111	4.655

Source: Metcalf and Eddy, Inc. 1975. Report to the National Commission on Water Quality: Assessment of Technologies and Costs for Publicly Owned Treatment works under P.L. 92-500. PB 250 690. Springfield, Va.: National Technical Information Service.

TABLE C.5

Costs of Components Not Included in Cost Functions, by Flow
(millions of 1973 dollars)

Component	Flow (millions of gallons per day)			
	.1	1.0	5.0	10.0
Miscellaneous structures and sitework	.030	.080	.110	.120
Outfalls	.010	.040	.105	.113
Pumping	.200	.350	.700	1.000
Total	.240	.470	.915	1.233
percent of treatment	61	37	30	26

Source: Metcalf and Eddy, Inc. 1975. Report to the National Commission on Water Quality: Assessment of Technologies and Costs for Publicly Owned Treatment works under P.L. 92-500. PB 250 690. Springfield, Va.: National Technical Information Service.

TABLE C.6

Sewer Cost Estimating Factors

Population Size Range	Collector Sewer Capital Costs (June 1973 dollars)	Interceptor Sewer Factor (multiplied by treatment plant cost)
0–500	580	.373
500–1,000	479	.384
1,000–2,500	406	.498
2,500–5,000	348	.722
5,000–10,000	319	.790
10,000–25,000	274	1.060
25,000–50,000	244	1.487
50,000–100,000	210	1.583
100,000–250,000	178	1.763
>250,000	129	2.473

Sources: Collector costs derived from Metcalf and Eddy, Inc. 1975. Report to the National Commission on Water Quality: Assessment of Technologies and Costs for Publicly Owned Treatment Works Under P.L. 92-500. PB 250 690. Springfield, Va.: National Technical Information Service; and interception factors from Michel, Robert L. 1970. Factors Affecting Construction Costs of Municipal Sewer Projects, mimeographed.

Adjustments of Capital Cost Estimates

Total capital costs for industrial and municipal activities, based on the capital cost functions described previously, were adjusted to take into account two factors. One adjustment was necessary to approximate the total capital investment required to build an operational facility. The other adjustment was to convert our capital cost estimates from 1973 dollars to 1975 dollars in order to be comparable to NCWQ.

In order to approximate the total capital investment required to build municipal sewerage treatment systems, we included the cost of building interceptor sewers as well as the costs of treatment. EPA and professional engineers have viewed interceptor sewers as an integral part of a treatment system because of the necessity to transport the wastes in collector sewers to sewerage treatment plans. In the analysis of municipal costs, we estimated that interceptor sewer costs were approximately 40 percent of the treatment costs. Consequently, we increased the estimated capital costs of residual reduction technology for municipal activities by 40 percent in order to take into account the integral relationship between treatment and interceptor sewers.

In order to approximate the total capital investment required to build industrial end-of-pipe treatment technology, we included adjustments for essential unit processes not included in the cost estimates and for a more realistic estimate of contingency and all other factor charges. The essential unit processes not included in the industrial capital estimate were for pumping, outfalls, and miscellaneous structures and sitework. These unit processes add on average an additional 40 percent to the costs of treatment and sludge disposal facilities (see Tables C.4 and C.5). An adjustment for a more realistic estimate of contingency was assumed to be 5 percent, and for all other factors was assumed to be another 5 percent. Consequently, we increased the estimated capital costs of residuals reduction technology for all industrial activities by 40 percent in order to compensate for unit processes not considered and an underestimate of contingency and other factor costs.

The other major adjustment was to convert our capital cost estimates from 1973 to 1975 dollars. Since the NCWQ converted their technology capital cost estimates in 1973 dollars to 1975 dollars by a factor of 1.3, we used the same factor to be consistent with their estimates.

In summary we adjusted our total capital cost estimates in order to compensate for technologies not included in the treatment costs, to correct for underestimates of contingency and other factors, and to convert to 1975 dollars. This adjustment factor of 1.82 (100 x 1.4 x 1.3) was applied to all BPT/ST and BAT/BPWTT capital cost estimates.

Sewer costs used in the municipal segment of NRDI are presented in Table C.6.

REFERENCES

Associated Air and Water Resources Engineers. 1973. Estimating Water Pollution Control Costs from Selected Manufacturing Industries in the U.S. 1973-1977. Washington, D.C.: Environmental Protection Agency.

Bechtel, Inc. 1975. Cost Effective Wastewater Treatment Systems. Washington, D.C.: Environmental Protection Agency.

Datagraphics Inc. 1975. Wastewater Treatment Process Cost Functions. Carnegie, Pa., mimeographed.

Environmental Protection Agency. 1973. Process Design Manual for
    Carbon Adsorption. Washington, D.C.: Environmental Protec-
    tion Agency.
    _____. 1975a.Development Document for Final Effluent Limitations
    for Fruits, Vegetables, and Specialties. Washington, D.C.:
    Government Printing Office.
    _____. 1975b. Development Document for Interim, Final, and Proposed
    Effluent Limitation Guidelines for Ore Mining and Dressing
    Industries. Vol. 2. Washington, D.C.: Government Printing
    Office.
Metcalf and Eddy, Inc. 1975. Report to the National Commission on
    Water Quality: Assessment of Technologies and Costs for Publicly
    Owned Treatment Works Under P.L. 92-500. PB 250 690.
    Springfield, Va.: National Technical Information Service.
Michel, Robert L. 1970. Factors Affecting Construction Costs of
    Municipal Sewer Projects, mimeographed.
National Research Council. 1975. National Residuals Discharge
    Inventory Computer Outputs. Washington, D.C.: National
    Research Council.

RALPH A. LUKEN is an international economics consultant currently based in Asia. Some of his more recent engagements have been with the United National Development Program and the Government of Thailand.

Dr. Luken has Ph.D's in Economics (1974) and Natural Resource Planning (1967) from the University of Michigan. From 1971 to 1974, Dr. Luken was employed by the U.S. Environmental Protection Agency. He evaluated most aspects of the implementation of the 1972 Water Act and served as chairman of working groups on the Environmental Financing Authority and the Alternate Financing Study. Prior to that, he also spent time as an Assistant Professor at the University of Michigan and as a Post-Doctoral Fellow at Johns Hopkins University.

Dr. Luken began his independent consulting practice in 1974. His clients have included the National Commission on Water Quality, The Old West Regional Commission, and the National Academy of Sciences/National Research Council.

Dr. Luken is the author of several papers and reports, was editor of the 1973 Economics of Clean Water, wrote a chapter in Environmental Quality and Water Development, and is the author of Preservation Versus Development: An Economic Analysis of San Francisco Bay Wetlands.

EDWARD H. PECHAN, III is a Senior Environmental Analyst with the Office of Planning, Analysis, and Evaluation of the U.S. Energy Research and Development Administration. He is responsible for developing methodologies to evaluate environmental impacts expected to result from emplacement of future energy systems.

After graduating from the University of Pittsburgh with a B.S. in Physics, Mr. Pechan served two years as a principal in a computer software development firm. He received his M.S. in Operations Research in 1970 from the University of Pittsburgh and spent two years as a Program Manager for a small consulting firm. In late 1972 he moved to Washington, D.C. to join the U.S. Environmental Protection Agency as a consultant and operations research analyst. During this period, Mr. Pechan led several efforts to improve accuracy and usefulness of environmental information.

During most of 1975, Mr. Pechan served as a consultant to the National Academy of Sciences/National Research Council. It was during this period that much of the material for this book was developed.

Mr. Pechan has authored numerous publications including several papers on air- and water-pollution modeling and a chapter in Models for Environmental Pollution Control.

CLEANING UP EUROPE'S WATERS: Economics, Management, and Policies

> Ralph W. Johnson
> Gardner M. Brown, Jr.

ENVIRONMENTAL LEGISLATION: A Sourcebook

> edited by
> Mary Robinson Sive

GLOSSARY OF THE ENVIRONMENT

> edited and translated by
> Paul Brace

MANAGING ENVIRONMENTAL CHANGE: A Legal and Behavioral Perspective

> Joseph F. DiMento

MANAGING SOLID WASTES: Economics, Technology, and Institutions

> Haynes C. Goddard

PRESERVATION VERSUS DEVELOPMENT: An Economic Analysis of San Francisco Bay Wetlands

> Ralph Andrew Luken